CHRISTIANITY AND SCIENCE

THEOLOGY IN GLOBAL PERSPECTIVE SERIES

Peter C. Phan, General Editor
Ignacio Ellacuría Professor of Catholic Social Thought,
Georgetown University

At the beginning of a new millennium, the *Theology in Global Perspective* Series responds to the challenge to re-examine the foundational and doctrinal themes of Christianity in light of the new global reality. While traditional Catholic theology has assumed an essentially European or Western point of view, *Theology in Global Perspective* takes account of insights and experience of churches in Africa, Asia, Latin America, Oceania, as well as from Europe and North America. Noting the pervasiveness of changes brought about by science and technologies, and growing concerns about the sustainability of Earth, it seeks to embody insights from studies in these areas as well.

Though rooted in the Catholic tradition, volumes in the series are written with an eye to the ecumenical implications of Protestant, Orthodox, and Pentecostal theologies for Catholicism, and vice versa. In addition, authors will explore insights from other religious traditions with the potential to enrich Christian theology and self-understanding.

Books in this series will provide reliable introductions to the major theological topics, tracing their roots in Scripture and their development in later tradition, exploring when possible the implications of new thinking on gender and socio-cultural identities. And they will relate these themes to the challenges confronting the peoples of the world in the wake of globalization, particularly the implications of Christian faith for justice, peace, and the integrity of creation.

Other Books Published in the Series

Orders and Ministries: Leadership in a Global Church, Kenan Osborne, O.F.M.
Trinity: Nexus of the Mysteries of Christian Faith, Anne Hunt
Eschatology: The Language of Hope, Anthony Kelly, C.Ss.R.
Meeting Mystery: Liturgy, Worship, Sacraments, Nathan D. Mitchell
Creation, Redemption, Grace, Neil Ormerod
Globalization, Spirituality, and Justice: Navigating a Path to Peace, Daniel G. Groody

THEOLOGY IN GLOBAL PERSPECTIVE SERIES
Peter C. Phan, General Editor

CHRISTIANITY AND SCIENCE

Toward a Theology of Nature

JOHN F. HAUGHT

ORBIS BOOKS

Maryknoll, New York 10545

Founded in 1970, Orbis Books endeavors to publish works that enlighten the mind, nourish the spirit, and challenge the conscience. The publishing arm of the Maryknoll Fathers and Brothers, Orbis seeks to explore the global dimensions of the Christian faith and mission, to invite dialogue with diverse cultures and religious traditions, and to serve the cause of reconciliation and peace. The books published reflect the opinions of their authors and are not meant to represent the official position of the Maryknoll Society. To obtain more information about Maryknoll and Orbis Books, please visit our website at www.maryknoll.org.

Library of Congress Cataloguing in Publication Data

Haught, John F.
    Christianity and science : toward a theology of nature / John F. Haught.
        p. cm. — (Theology in global perspective)
    Includes index.
    ISBN 978-1-57075-740-2
    1. Nature—Religious aspects—Christianity. I. Title.
    BT695.5.H37 2007
    261.5'5—dc22
                                2007005355

Contents

Foreword

by Peter C. Phan

The relationship between science and Christianity, to put it in a nutshell, has been one of love and hate. Like a divorced couple who want to restore their marriage, Christianity and science—Christianity at any rate—may want to remember the time of romance when harmony reigned supreme and collaboration promised to be a wonderful common adventure. But many obstacles and misunderstandings caused their breakup. Out of this painful memory and a willingness to change, perhaps new possibilities will emerge for a more enriching relationship, or at least, for overcoming rancor and antagonism.

The relationship between science and Christianity is part of, though not reducible to, the larger story of the interaction between reason and faith, or between philosophy and theology. Again, like partners in marriage, the two protagonists have not remained unchanged, nor has the dynamic of their relationship. On the one hand, Christianity is not a monolith; in fact, it is historically more correct to speak of Christianities, even in the West. Moreover, the approaches of Christians to reason, philosophy, and secular knowledge have been extremely varied, ranging from admiration and creative assimilation, for example, in Origen and Thomas Aquinas, to rejection and condemnation as championed by Tertullian and Luther. On the other hand, science itself is not a homogenous branch of knowledge with a single object and a uniform method of investigation. "Science" includes such widely divergent disciplines as biology, physics, psychology, and sociology—just to mention a few that have presented severe challenges to Christianity—and their methodologies have undergone vast transformations. Many practitioners of the so-called hard sciences have largely abandoned a positivistic worldview and a strictly empiricist methodology and have begun investigation of areas that lie beyond the domain of empirical verification. This does not mean that scientists and Christians have entered into an amicable relationship. Rather, the relationship has been very checkered, as anyone with a passable knowledge of Galileo Galilei, Karl Marx, Charles Darwin, and Sigmund Freud will understand.

The book that lies before you will tell one part of the long story of the relationship between science and Christianity, with the focus on contemporary cosmology. John Haught, whose writings on this theme have earned him an international reputation, asks us to revisit Christian theology while keeping in mind the scientific discovery that the universe is an evolving story. If our human history takes up only the last few lines of the last volume of a thirty-volume set, each volume 450 pages long, and if this set is only a part of a huge library, then, Haught asks, how are we going to understand "ourselves, God, creation, Trinity, Christ, redemption, incarnation, faith, hope and love," in the light of the three "infinites" of the universe, that is, infinitely big, infinitely small, and infinitely complex?

In helping us find the answer, Haught takes his cue from the French Jesuit paleontologist and theologian, Pierre Teilhard de Chardin, who was sometimes in conflict with his ecclesiastical superiors for his ideas on Christian faith and evolution. In ten lucid chapters, Haught reformulates the Christian answers to the questions he raises, always keeping before his eyes the fact that the universe—or more precisely, the "multiverse"—is evolving. It is of great significance that he begins his exposition with meditations on hope and mystery, without which neither Christian life nor science would make sense.

As Haught moves to consider Christian beliefs such as revelation, creation, divine providence, incarnation, the divine Spirit, death, and resurrection, he proves to be a sure and trustworthy guide in helping us see that "there is nothing in Christian faith that should make one afraid of science's widening and deepening of knowledge . . . The more extended and elaborate our sense of creation becomes, the more we should be able to enlarge our appreciation of the world's Creator and the scope of divine purpose and providence. Science may be offering us, therefore, not less but more reason than ever for worship and gratitude."

The marriage between science and Christianity, then, is neither an arranged marriage nor a marriage of convenience. If this marriage has proved rocky in the past, Haught shows how it could be repaired, with each partner, as we say in Vietnamese, *nhan loi* (acknowledging one's faults), *xin loi* (asking for forgiveness of one's faults), and *sua loi* (correcting one's faults). The repair of this marriage between science and Christianity is worth the best efforts of both scientists and theologians, since neither science nor religion can reach its full potential without the other.

Preface

One of the most surprising scientific discoveries of the past century and a half is that the universe is an unfolding story.[1] The sense that the universe is still in the process of coming into being began to emerge faintly several hundred years ago when Tycho Brahe and Galileo Galilei produced visual evidence that the heavens are not changeless. Today, however, developments in geology, evolutionary biology, and cosmology have left no doubt: the whole of nature, not just earth and human history, has an essentially narrative character. Before modern times the wider universe seemed to be the general context and container of local, terrestrial stories, but not itself a story. Now science has shown that our universe is still undergoing transformations that can best be represented in the style of a drama. Formerly the heavens seemed steady enough to frame all the stories unfolding on earth. The firmament was a place of refuge to which worldlings could flee, at least in contemplation, from the fatal flow of events here below. But during the last century the heavens too were swallowed up by a story, one that now seems almost too large for the telling.

What is Christian theology going to make of this larger story? The unfathomable reach of cosmic proceedings infinitely outstrips in time and space the brief span of human flourishing and the even more fleeting moments of Hebrew and Christian religious history. Science has discovered a world that moves on a scale unimaginable to the prophets and evangelists. Is it possible that the universe has outgrown the biblical God who is said to be its Creator? Many thoughtful people today have concluded that this is exactly what has happened. The very substance of Christian faith seems irreversibly intertwined with the outworn imagery of an unmoving planet nested in an unchanging cosmos. Pictures of nature that had been fixed in the minds and feelings of people for centuries prior to the birth of science need to be redrawn. But can this occur without a radical revision of faith and theology?

Can Christianity and its theological interpretations find a fresh foothold in the immense and mobile universe of contemporary science, or will science itself replace our inherited spiritualities altogether, as many now see happening? The Jesuit geologist Pierre Teilhard de Chardin asks: "Is the Christ of the Gospels, imagined and loved within the dimensions of a Mediterranean world, capable

---

1. Carl Friedrich von Weizsäcker, *The History of Nature* (Chicago: University of Chicago Press, 1949); see also Stephen Toulmin and June Goodfield, *The Discovery of Time* (London: Hutchinson, 1965); Wolfhart Pannenberg, *Toward a Theology of Nature: Essays on Science and Faith*, ed. Ted Peters (Louisville: Westminster John Knox, 1993), 86-98.

of still embracing and still forming the centre of our prodigiously expanded universe?"[2] Is it not the case that science has changed things too fast for Christianity and other faiths ever to catch up? Isn't it time for all people to wake up from their religious slumbers and bind themselves to the more elegant creed of pure naturalism? Isn't nature itself now sufficiently immense to satisfy human longing for infinite mystery? And isn't science a more trustworthy guide than theology as we venture forth into nature's newly discovered depths?[3]

Before beginning a response to these questions, let us first try to get a visual impression of the universe's vast dimensions as science sees them today. Imagine that you have thirty large volumes on your bookshelf. Each tome is 450 pages long, and every page stands for one million years. Let this set of books represent the scientific story of our 13.7-billion-year-old universe. The narrative begins with the Big Bang on page 1 of volume 1, but the first twenty-one books show no obvious signs of life at all. The earth story begins only in volume 21, 4.5 billion years ago, but life doesn't appear until volume 22, about 3.8 billion years ago. Even then, living organisms do not become particularly interesting, at least in human terms, until almost the end of volume 29. There the famous Cambrian explosion occurs, and the patterns of life suddenly burst out into an unprecedented array of complexity and diversity. Dinosaurs come in around the middle of volume 30 but are wiped out on page 385. Only during the last sixty-five pages of volume 30 does mammalian life begin to flourish. Our hominid ancestors show up several pages from the end of volume 30, but modern humans do not appear until the bottom of the final page. The entire history of human intelligence, ethics, religious aspiration, and scientific discovery takes up only the last few lines on the last page of the last volume.

After scanning this set of thirty volumes, try now to conjure up a whole *library* of such sets, with archives extending indefinitely in all directions. Scientists today are finding it increasingly plausible that our Big Bang universe is only one of countless many worlds. The thirty volumes on your shelf are a mere outpost in an infinitely larger *multiverse*. It is on such a scale that much scientific inquiry is beginning to move today. Can theology keep pace?

Next, turn to a single dot on any page of this book and enter through that portal into the world of the unimaginably small. As you journey deeper into that invisible realm conjure up the reverse image of worlds within worlds, now too small and too subtle to be represented pictorially in three paltry dimen-

---

2. Pierre Teilhard de Chardin, *The Divine Milieu: An Essay on the Interior Life* (New York: Harper & Row, 1960), 46.

3. Ursula Goodenough, *The Sacred Depths of Nature* (New York: Oxford University Press, 1998); Chet Raymo, *Skeptics and True Believers: The Exhilarating Connection between Science and Religion* (New York: Walker, 1998).

sions. In the direction of both the large and the small, science has now rendered our old maps obsolete. There are at least two "infinites" that catch our attention today, Teilhard observes, one of the *immense*, the other of the *infinitesimal*.[4] But there is also a third, one that gets less press than the other two. It is the infinite of *complexity*. In the sphere of living and thinking beings, for example, the particles of physics and the elements of chemistry have been taken up into emergent cells and organisms wherein they are patterned so intricately that all attempts to isolate and specify the individual roles of the component units are met with frustration. We may also call this intricate patterning the infinite of *relationality*. In a cell or organism—especially those endowed with nervous systems and brains—every component is so interior to, and constitutive of, the identity of every other that we cannot understand an organism by taking it apart. If we dissect it, we murder it. An organism is a bundle of connections that interlace, overlap, and feed back into one another in endless dynamic interplay. To isolate any part of this network is to miss its meaning altogether.

What are faith and theology to make of a world embedded in the three infinites—the immense, the infinitesimal, and the complex—and the picture of nature they entail? In Pascalian terms, how are we Christians to understand ourselves amid the three infinites that science has opened up to our awestruck sensibilities? Now that we find ourselves webbed into an unimaginable cosmic tapestry and an unfathomable temporal depth and spatial extension, what will it mean for our understanding of ourselves, God, creation, Trinity, Christ, redemption, incarnation, faith, hope, and love?

I believe there are three general ways of responding to this question. First, one may go on pretending that science has never happened or that it is talking about things completely unrelated to faith and theology. Accordingly, there is no need to make any religious or theological readjustments in the face of new knowledge of the three infinites. Theologians may have to demythologize their holy books so that people will no longer confuse ancient cosmology with the religious content buried in the texts. But at all costs the substance of faith must be protected from taking on radically new meaning as a result of the changing spheres of scientific discovery.

A second response is to discard faith and theology altogether as parasitic riders on now obsolete cosmologies. According to those who adopt this response—I shall be calling them "scientific naturalists"—classic religious beliefs are so dependent on outmoded and undersized cosmologies that the

---

4. See Pierre Teilhard de Chardin, *The Human Phenomenon*, trans. Sarah Appleton-Weber (1959; Portland, Ore.: Sussex Academic Press, 1999), 217.

ancient creeds have already begun to evaporate in the noonday sun of scientific enlightenment. Theology, along with its attendant cosmologies, can survive only so long as people remain ignorant of what science has now revealed.

Third, one can embrace the three infinites, or, better, be embraced by them, in such a way as to read them as invitations to an unprecedented magnification of the sense of God, creation, Christ, and redemption. I propose that we try out this latter approach. There is nothing in Christian faith that should make one afraid of science's widening and deepening of knowledge. No matter how enormous the picture of the natural world turns out to be, it can never surpass the infinity we have always attributed to God. The more extended and elaborate our sense of creation becomes, the more we should be able to enlarge our appreciation of the world's Creator and the scope of divine purpose and providence. Science may be offering us, therefore, not less but more reason than ever for worship and gratitude.

METHOD

This is a work of systematic theology, an enterprise animated by faith but structured by reason. Reason, as I understand it, is inevitably abstract, but abstractions are necessary in order to focus our finite minds in ways that allow the world to become intelligible. Different systematic theologians focus their study of the religious meaning of phenomena in different ways, of course, and this means that each theological program will be only a glimpse of all that needs to be said. Every theological system will require criticism and supplementing from those that focus their content in other ways. Consequently, I confess at the very outset that the outline of a theology of nature laid out in the following pages can illuminate only a small patch. I make no attempt to be comprehensive. Nor do I see this work as a substitute for historical studies of science and theology. Instead, I shall be examining some of the discoveries of natural science, especially physics, biology, and cosmology, and ask what they might mean for Christian faith. In order to give coherence and structural constraints to this effort I have decided to look at modern and recent scientific understanding of the natural world from the point of view of two closely related motifs of Christian faith: the *descent* and *futurity* of God. What these concepts mean will be developed as I proceed. Here I want only to make very clear the limitations of this work.

I began developing the theological framework of the following reflections on science and nature some years ago in a work on the theology of revelation.[5]

---

5. John F. Haught, *Mystery and Promise: A Theology of Revelation* (Collegeville, Minn.: Liturgical Press, 1993).

1

Science and Christian Hope

Shatter, my God, through the daring of your revelation the childishly timid outlook that can conceive of nothing greater or more vital in the world than the pitiable perfection of our human organism.
—Teilhard de Chardin[1]

CHRISTIAN FAITH IS ESSENTIALLY about the future—not simply a future beyond the world but also the future *of* the world. Of course, it is concerned also with the meaning of the present and the past, but what this meaning is can be fully revealed only in the future. Christian faith is above all a quest for the Ultimately New, and it hopes for the radical renewal of "the whole of reality," not just human history.[2] Christians are called upon to extend their religious expectations beyond human preoccupations outward toward the entire universe and its future. Science can help them do so.

Science, as I noted in the preface, has exposed the three infinites—the immense, the infinitesimal, and the complex. But Christian faith had already opened up a fourth, the infinite horizon of the future. It is the Future beyond all futures that Christian hope seeks. The heavens may entrance us, but even in their staggering expansiveness we cannot find all that our hearts are longing for. The human spirit's quest for final liberation leads beyond all present times and past all perishing, beyond this universe and any others, toward the Absolutely New—in other words, to God, the one whose promises open up all of life and all universes to an endless and unimaginable future. "Christian hope," says theologian Jürgen Moltmann, "is directed towards . . . a new creation of all things by the God of the resurrection of Jesus Christ."[3] At the very heart of Christianity lies a trust that the world remains forever open to a new future. The name of this future is "God." God, however, is not just any future

---

1. Pierre Teilhard de Chardin, "The Mass on the World," in Thomas M. King, *Teilhard's Mass: Approaches to "The Mass on the World"* (New York: Paulist, 2005), 150.

2. Jürgen Moltmann, *Theology of Hope: On the Ground and the Implications of a Christian Eschatology,* trans. James W. Leitch (New York: Harper & Row, 1967), 34.

3. Ibid., 33.

that we dream up. The futures we conjure up and plan for ourselves are inevitably inadequate to what we really need. Rather, as Karl Rahner has put it, God is the *Absolute Future*, deeper and more surprising than anything we could possibly wish for ourselves. *Deus semper maior* ("God is always greater").[4]

God is the "power of the future"[5] that rises up to greet the universe anew at the place where each present moment passes away. Although we cannot grasp this elusive Future we can allow ourselves to be grasped by it. "The coming order is always coming, shaking this order, fighting with it, conquering it and conquered by it. The coming order is always at hand. But one can never say: 'It is here! It is there!' One can never grasp it. But one can be grasped by it."[6]

Perhaps "future" is not the first idea that people today, including Christians, associate with the word "God." The essential "futurity" of God that shaped the biblical experience has for centuries hidden behind a fogbank that is only now, very slowly, beginning to dissipate. As the mists that had enshrouded the future begin to fade, we may still prefer not to expose ourselves to the wide vista that opens up ahead. The future that Israel, Jesus, and the early church felt to be dawning so dramatically, the "coming of God" that gave their lives a sense of adventure and unparalleled excitement, many of us would still prefer to hold at bay. The restlessness that accompanies exposure to the future is easily suppressed, especially if we are comfortable with the way things are right now.

And yet, even in the best of circumstances, at some level of our being we still long for a new future, even as we cling to what is past or present. A sense of the coming (*adventus*) of God stirs us up, makes us yearn for deeper freedom, for a more wide-open space in which to live. Yet, like idlers standing by in the market place, we remain tied to what is or what has been rather than to what will be. It is the destitute, those who now have nothing to fall back on, who are most open to the promise of a radically new world. It is their ears that the fire of the gospel first singes with the unsettling news of God's coming.

---

4. Karl Rahner, *Theological Investigation*, vol. 6, trans. Karl and Boniface Kruger (Baltimore: Helicon, 1969), 59-68; see also Jürgen Moltmann, *The Experiment Hope*, ed. and trans. M. Douglas Meeks (Philadelphia: Fortress, 1975), 48.

5. Wolfhart Pannenberg, *Faith and Reality*, trans. John Maxwell (Philadelphia: Westminster, 1977), 58-59; Ted Peters, *God—The World's Future: Systematic Theology for a New Era*, 2nd ed. (Minneapolis: Fortress, 2000).

6. Paul Tillich, *The Shaking of the Foundations* (1948; New York: Charles Scribner's Sons, 1996), 27.

But how are we to connect the thought world of the natural sciences to the Christian revelation of a God who is coming and seeks to renew the world? If we are receptive to the gospel and serious about making sense of Christian faith today, we need to tie what science is telling us about the universe to Jesus' own infectious excitement about the coming of God's reign. The fervor of expectation aroused in his followers by Jesus and the news of his resurrection must be the framework of any truly Christian reflection today on the meaning of the entire universe as it is being laid open to us by the natural sciences. A Christianity that avoids reflecting on scientific understanding of what is going on in the universe is less than realistic. Christian faith needs to be not only consistent with what the sciences are saying but also eager to render more intelligible than ever the world that science has been setting before us. It is the purpose of this book to suggest ways in which science can influence and challenge Christian faith as well as how the light of faith can illuminate what we are learning from science about nature.

SCIENTIFIC SIMPLIFICATION

Can one worship the Christian God in an age of science? For many it is not easy. On the one hand, scientific *discovery* has made the universe appear larger and more complex than ever before—for many even larger than God. On the other hand, scientific *method* seems, at least at first sight, to make the world smaller than it really is. It can make the universe seem too simple to inspire a sense of mystery. Science's way of investigating is to break natural phenomena down into more elementary components or into earlier chains of physical causes. Likewise, scientific method deliberately blinds itself to what Pierre Teilhard de Chardin calls the "insideness" of things.[7] Science looks at the world externally and objectively. It says nothing about value or meaning, and it misses completely the subjective world that each of us experiences interiorly. Moreover, science studies events in terms of *what has been* rather than what will be. Of course, it tries to make the future predictable, for unless science can make predictions, it does not qualify as science. But it can predict what will happen in the future only on the basis of what has already happened. By itself scientific method leaves little room for *new* being. On its own it is scarcely able to hear the strains of any dawning new creation. As I shall emphasize repeatedly, in harmony with one of Teilhard's most important

---

7. Pierre Teilhard de Chardin, *The Human Phenomenon*, trans. Sarah Appleton-Weber (1959; Portland, Ore.: Sussex Academic Press, 1999), 23-24.

principles, the world can become fully intelligible to us only as we look toward its future, not by looking only into its historical past or particulate makeup.[8]

Scientific method then is unable by itself to prepare the mind and heart for what is truly new. But Christian faith, as distinct from science, is *essentially* expectation of new creation, so its prescribed posture of hope may sometimes seem remote from the alleged "realism" of science. Christian creeds, doctrines, and theologies are not properly read unless they communicate the sense of expectation that gave rise to the faith's earliest fervor about the coming of God. "We hope to enjoy forever the vision of Your Glory," many of us pray during the eucharistic celebration, but how can we express such hope and at the same time accept what the sciences are telling us about the world? Christian hope implies that the world is not ultimately tied to endless repetition; yet in science everything must conform to timeless, hidebound routine. How then can we hold science and faith together without contradiction? This, it seems to me is a central question for theology today, and it will not go away simply by our ignoring it.

Unfortunately, Christian instruction can also easily anesthetize human minds against the invasion of the future. Theology has often represented the idea of God in concepts that deal best with what is, or what has been, rather than with what will be. God is usually pictured as the eternal, unchanging, and timeless mystery that grounds, creates, and now hovers over, or underlies, the world. Many Christians have grown comfortable with such vertical locations of deity, but these depictions of God, supported as they are by prescientific theological metaphysics, fail to capture the mood of expectation that haunted the earliest ecclesial gatherings.

It is doubtful that theology can be faithful to its calling as long as it fails to put us in touch once again with the *anticipatory* temper of Christian faith. But it is precisely the expectation of new creation that makes it so difficult for many scientists and philosophers to accept Christianity. Of course, there are also less important—and entirely unnecessary—reasons why the scientifically enlightened often disdain Christianity, as well as why many Christians snub science. We shall have ample opportunity to review these reasons later on. The point I want to make here is that for many scientifically educated people the *true* stumbling block to Christian faith is its belief that a new world is coming and indeed is already taking hold right now, transforming and renewing the *whole* of creation.

Scientific method is simply not equipped to see this happening. Science views the present in terms of what is earlier and simpler. Its sense of the future

---

8. Ibid., 163: It is only in the future that "the past lines of evolution take on their maximum coherence."

is shaped by a preoccupation with what has already taken place in strict conformity with the virtually timeless laws of physics and chemistry. Science, in other words, is not wired to see what is truly new. Genuine Christian faith, on the other hand, views things and events especially in terms of what is coming. Science is not wrong in looking to the past so as to understand the present, but its way of seeing the world is limited. If the world has room for a radically new future, scientific method is not perceptive enough to take hold of it. Christian faith, as I shall emphasize throughout, is essentially about what is coming, and about the God whose very essence is to be *future*, the inexhaustible font of renewal.[9]

Does this mean then that science and faith are incompatible ways of looking at the world? Not at all. Not only are they compatible, but a mutual engagement of the two perspectives can enrich the lives of all of us. Looking with science toward what is earlier-and-simpler is essential to appreciating the arrival of what is later-and-more. And a sense of the dimly dawning horizon of a later-and-more can give deeper meaning to what science sees in its survey of the past and present. This mutuality will prove especially significant as we try to understand the phenomena of emergence and evolution.

Science looks at the world by observing and generalizing from large numbers of similar events that have already taken place. Every falling object, for example, traces the same changeless path of acceleration that the Newtonian law of gravity specified several centuries ago. Every new species of life can be accounted for by looking back at how the invariant mechanism of natural selection has eliminated unpromising traits of organisms in the past. In this sense science abides no exceptions and no surprises. Scientific method focuses on initial physical conditions and the unchanging, deterministic laws operating in nature from age to age. Of course, it may occasionally discover habits of nature previously unknown, and it can formulate fresh hypotheses. But, at least as it has been understood for the past several centuries, science looks at things in terms of what has always been. To science, every future occurrence, no matter how strange, will be an exemplification of timeless laws and previous physical circumstances, so by itself science simply cannot see clearly the perpetual newness of creation. Even though dramatically new phases, or new kinds of physical activity, such as living and thinking organisms, have emerged at times in nature's long history, science tries to explain these "emergent" phenomena as much as possible in terms of the earlier habits of nature pertaining to nonliving and unintelligent physical processes.

---

9. Following Jürgen Moltmann, *The Coming of God: Christian Eschatology*, trans. Margaret Kohl (Minneapolis: Fortress, 1996).

There is irony here, for the actual *discoveries* of science, such as the Big Bang, the evolutionary trajectory of life, the genetic code, the Hubble Deep Field, and the chemical aspects of mind have in fact made the world new for all of us. In this sense science continually opens up a new future to human consciousness. Scientists, as human persons just like the rest of us, face their professional future in expectation of arriving at new insights. This anticipation energizes them and gives meaning to their lives. But when it comes to *explaining* new discoveries, science is typically limited to fitting them into what it already knows of nature's past patterns of occurrence.

Even new theories, at least until fresh information challenges the old, are made to fit into an established understanding of the laws of physics, or else they are not scientifically intelligible. Unless natural phenomena and processes can be simplified in such a way as to be mathematically intelligible, our understanding of them will not qualify as scientific. As the mathematician Gregory Chaitin puts it, "for any given series of observations there are always several competing theories, and the scientist must choose among them. The model demands that the smallest algorithm, the one consisting of the fewest bits, be selected. Put another way, this rule is the familiar formulation of Occam's razor: Given differing theories of apparently equal merit, the simplest is to be preferred."[10]

This reductive approach, once again, is not wrong. Methodologically speaking, science has every right to look at the world in a way that momentarily brackets out the impression of novelty. Much can be learned about nature by focusing in mathematical terms on the regularities it always obeys. And scientific explanation must be a legitimate part of any rich accounting for everything that goes on in nature, even intellectual, moral, and religious activity. The salient issue, however, is whether science can be the *whole* explanation.

In the intellectual world today, apart from scattered islands of postmodern dissent, there is a widely shared belief that science is enough to account fully for everything. "Scientific naturalism," as I shall call it, is the academically endorsed belief that science alone can take us down to the deepest and most fundamental strata of the world's being.[11] So it is not scientific method itself but *belief* in the unlimited explanatory scope of science that contradicts Chris-

---

10. Gregory J. Chaitin, "Randomness and Mathematical Proof," in *From Complexity to Life: On the Emergence of Life and Meaning,* ed. Niels Gregersen (New York: Oxford University Press, 2003), 23.

11. Apparently it was T. H. Huxley, Charles Darwin's famous "bulldog," who first used the expression "scientific naturalism." See Ronald Numbers, "Science without God: Natural Laws and Christian Belief," in *When Science and Christianity Meet,* ed. David C. Lindberg and Ronald Numbers (Chicago: University of Chicago Press, 2003), 266.

tianity and other religions. To Christian theology the sciences are important levels in an extensively layered hierarchy of explanations needed to account for anything. But science, since it leaves so much of the world out of its theories, hypotheses, and mathematical models, cannot provide ultimate explanations.

Theology, on the other hand, professes to instruct people about the deepest explanatory level of all. It looks for ultimate explanation, whereas science is limited to proximate ones. To theology, the ultimate explanation of nature and nature's laws is the creativity, love, power, and wisdom of God, the One who perpetually opens up a new future for the world. Scientific naturalism, on the other hand, assumes that science, at least in principle, can explain all things exhaustively and ultimately in terms of what has already been. In its belief that science alone can provide ultimate or final explanation, scientific naturalism in effect turns science into an alternative religion. And so it views theology as a rival rather than a friend of science.

If it is faithful to the biblical vision, theology for its part may find it necessary to cultivate what may be called a "metaphysics of the future."[12] From a Christian point of view the world leans on the future as its true foundation.[13] What gives consistency to the world—and happiness to the human heart—is the general thrust of all things toward what is yet to come. Were this forward momentum to slacken even momentarily, nature would be annihilated.[14] A richly textured understanding of the world, therefore, will not only uncover the past but also imagine its future. But such a forward thrust requires that our consciousness adopt the posture of anticipation, hope, and openness to surprise. Science alone, with its flair for tracing the world's journey back into the remote past, is not equipped to deliver this kind of discernment. Christianity invites us to look at the world through the eyes of hope. "From first to last," Jürgen Moltmann says, "Christianity . . . is hope, forward looking and forward moving, and therefore also revolutionizing and transforming the present." Hope is the "medium of Christian faith as such, the key in which everything in it is set, the glow that suffuses everything here in the dawn of an expected new day."[15]

---

12. I am using this expression in order to capture the biblical sense that what is "really real" for a community of hope lies "up ahead" in a future that has yet to be actualized by the God who is coming. For similar, but not identical, views, see Moltmann, *Experiment Hope*, 48; Pannenberg, *Faith and Reality*, 58-59; Rahner, *Theological Investigations*, 6:59-68. It is especially Rahner's understanding of God as the "Absolute Future" that lies behind my understanding here.

13. Pierre Teilhard de Chardin, *Activation of Energy*, trans. René Hague (New York: Harcourt Brace Jovanovich, 1970), 239.

14. Ibid.

15. Moltmann, *Theology of Hope*, 16.

A NEW DAY FOR THE UNIVERSE

The new day that Christianity expects, however, is not exclusively one of personal, political, and social liberation. It is also a new day for the entire universe, the heavens and the earth, for what is visible and invisible. And it is just this *cosmic* expectation on the part of Christianity that I shall emphasize in these pages. For Christian theology, Moltmann continues, there is essentially "only one problem: the problem of the future."[16] But the future will comprise not only those episodes of the human story that are yet to unfold but also the ongoing story of a still unfinished universe. If we fail to keep our sights trained on the distant cosmic future, and instead focus myopically only on human destiny, we shall shrink even our human hopes to the point where they no longer energize our lives and works. Consequently, this book's focus must be primarily on how the Christian hope for a new creation of the *cosmos* can frame the picture of the world that the natural sciences are now laying out before us.

Creeds that profess to be based on the biblical experience claim in effect that *everything* can be made new. In the modern period, however, there has emerged—not for the first time in human history—a pessimistic picture of the universe that denies that any *real* renewal of its being is possible. What seems new to us, this belief system maintains, is in fact always old and unchanging. There is really nothing new under the sun. The truly lucid consciousness, Albert Camus writes, must therefore be cleansed of hope.[17] Bertrand Russell echoes the sentiment: "Only on the firm foundation of unyielding despair can the soul's habitation henceforth be safely built."[18] And the Nobel Prize–winning physicist Steven Weinberg declares that "it would be wonderful to find in the laws of nature a plan prepared by a concerned creator in which human beings played some special role. I find sadness in doubting that we will."[19]

There can be no doubt that the birth of modern science, exciting as its discoveries have been, has simultaneously ushered in a fierce strain of pessimism about the future. Once again, this is partly because science looks essentially into the past, or into timeless laws of physics, for a "fundamental" under-

---

16. Ibid.

17. Albert Camus, *The Myth of Sisyphus, and Other Essays,* trans. Justin O'Brien (New York: Knopf, 1955).

18. Bertrand Russell, *Mysticism and Logic, and Other Essays* (1918; Garden City, N.Y.: Doubleday, 1957), 48.

19. Steven Weinberg, *Dreams of a Final Theory: The Scientist's Search for the Ultimate Laws of Nature* (New York: Pantheon Books, 1993), 256.

standing of how things will eventually turn out. Explanation, as far as natural science is concerned, means tracing a line of causation back into a series of events that have already happened. If you want to understand how snow rabbits came to be white, for example, you have to imagine a process of natural selection whereby predators *in the past* devoured all the dark or spotted rabbits who could not camouflage themselves against the snowy backdrop of a northern climate. And if you want to understand the expansion rate of the universe you have to go back fourteen billion years to the Big Bang itself.

Habituated to science's method of looking back, modern intellectual life has adopted a picture of reality that stands in tension with Christian hope and its expectation of future transformation. The practice of looking into a past inertial chain of causes to acquire present understanding has swept over the whole world. It has found a comfortable home in academic thought, and from there it has oozed out into modern and postmodern culture. It has shaped dominant views of economics, politics, and personality. It continues to influence social thought and the practice of medicine. It has even infiltrated the world of religious reflection. But its primary place of residence is the impressive edifice of the natural sciences.

This scientific abode has not only functioned as a forum in which to celebrate great discoveries and intellectual achievements, but it has also served as a kind of customs house, where all who enter into the world of "true" knowledge must check much of their cognitional apparel at the door. In exchange for a ticket to see what science has uncovered, visitors must agree not to ask questions about the meaning or value of things. They must look at the objects on display through lenses that filter out any shade of inherent importance or purpose. Further, they must focus not so much on wholes, but instead on component parts, processes, and mechanisms that cause things to function in a specific way.

For much of the modern period the search for explanation in the domain of what is earlier-and-simpler has meant endorsing the point of view known as "scientific materialism."[20] Materialism is the belief that reality consists ultimately of mindless and lifeless bits of "matter." This belief still provides the backdrop of much research. Today many philosophers call it "physicalism" rather than materialism in order to signify their awareness that during the past century matter has increasingly shown itself to be much more subtle and slippery than we used to think. But physicalism no less than materialism takes the natural world, as made accessible to us by science, to be all there is. The

---

20. For a profound analysis and critique of scientific materialism, see Alfred North Whitehead, *Science and the Modern World* (New York: Free Press, 1925), 51-59.

most fundamentally explanatory science, therefore, is physics. According to materialist philosopher David Papineau,

> physics, unlike the other special sciences, is complete, in the sense that all physical events are determined, or have their chances determined, by prior physical events according to physical laws. In other words, we never need to look beyond the realm of the physical in order to identify a set of antecedents which fixes the chances of subsequent physical occurrence. A purely physical specification, plus physical laws, will always suffice to tell us what is physically going to happen, insofar as that can be foretold at all.[21]

Materialism, or physicalism, implies a Godless world, whatever finer distinctions one might make. Here I wish only to indicate that in the intellectual world it is materialist belief, not science itself, that still constitutes the main challenge to religion and Christianity.

Any worldview that excludes the divine is also known more generally as "naturalism."[22] Naturalism is a broader notion than either materialism or physicalism, and it comes in many flavors. Its followers include not only the harder materialists and the softer physicalists but also those who are impressed by what seems to them to be the infinite resourcefulness and expansiveness of nature. Some naturalists are pantheists, others are "ecstatic naturalists," and still others are materialists. Some think the universe is for us, others against us. But, at least as I shall be using the term, naturalism is best defined as "the belief that nature is all there is."[23]

Naturalism arose historically—and understandably—as a reaction to a one-sided, world-despising supernaturalism, the kind of religiosity that finds in the transient natural world little to inspire hope and so looks for salvation only up above, in another world apart from this one.[24] In its extreme forms, supernaturalism blunts the sense of a future *for* the world, persistently translating the invigorating sense of the "up ahead" into a stagnating "up above," an interpretation that in turn leads at times to a religious hatred of nature. This perspective, it goes without saying, has little to do with the incarnational and eschatological perspective of biblical Christianity.

Naturalism is a powerful protest against extreme supernaturalism, and this protest comes in different forms. For example, there are both sunny and shady

---

21. David Papineau, *Philosophical Naturalism* (Cambridge, Mass.: Blackwell, 1993), 3.

22. Paul Tillich, *Systematic Theology*, 3 vols. (Chicago: University of Chicago Press, 1967), 2:5-10.

23. See Carl Sagan, *Cosmos* (New York: Ballantine Books, 1985), 1: "The universe is all that is, all there ever was and all there ever will be." Also Charley Hardwick, *Events of Grace: Naturalism, Existentialism, and Theology* (Cambridge: Cambridge University Press, 1996).

24. Tillich, *Systematic Theology*, 2:5-10.

naturalists. Sunny naturalists insist that nature is enough to satisfy all our spiritual needs. To them there is no need for traditional kinds of worship since the universe itself is large enough to fulfill our hearts' deepest longings. To the sunny naturalist, Christianity is misguided in focusing on a God distinct from nature. According to this species of naturalism the idea of God is not only scientifically unnecessary but also religiously and morally superfluous. Nature is enough. Shady naturalists, on the other hand, claim that since nature is the source of suffering and death, and not just of life and beauty, it would be silly to make any religious covenant with it. Shady naturalists are sad that the world seems so Godless. It would be comforting, they admit, to know that a beneficent providence governs the world, but scientific honesty requires that we now abandon such naïve trust. The world is headed toward final oblivion, so the best we can do is acquire a sense of honor for not having denied the fact of nature's tragic destiny.[25]

As far as the present study is concerned, I shall use the term "naturalism" to designate the broadly shared conviction, whether sunny or shady, that nature, as made available to ordinary experience and scientific discovery, is literally "all there is." And when I use the term "religion" I am referring henceforth to the belief that nature is *not* all there is. Christianity is a religion in this sense, but it is also one whose core teachings emphasize the goodness, and what I shall call the *promise,* of nature. In spite of the well-known historical difficulties associated with Galileo and Darwin, Christianity has no quarrel with science, as Pope John Paul II has recently emphasized.[26] But Christianity is inalterably opposed to naturalism. This book will take the position that it is not science but a kind of materialist naturalism often mistaken for science that stands in conflict with the beliefs of Christianity and other faiths. When Christian faith comes face to face with science itself, it finds a friend with whom it can converse, whatever blunders and misunderstanding there may have been in the past. With naturalism, however, there can be no fruitful coalition.[27]

Science, as Alfred North Whitehead acknowledged in the early twentieth century, has often metamorphosed into the defense of materialist naturalism,

---

25. Weinberg, *Dreams of a Final Theory,* 255, 260.

26. Pope John Paul II, "Address to the Pontifical Academy of Sciences," November 10, 1979; *Origins, CNS Documentary Service* 9, 24 (November 29, 1979), 391; also Cardinal Poupard, "Galileo: Report on Papal Commission Findings," *Origins* 22, 22 (November 12, 1992), 375; Pope John Paul II, "Letter to the Reverend George V. Coyne, S.J., Director of the Vatican Observatory," in *Origins* 18, 23 (November 17, 1988), 377.

27. This is not to say that Christians cannot form fruitful alliances with *naturalists* as distinct from naturalism. Indeed, on many ethical issues, such as the ecological predicament, cooperation is both warm and fertile.

much to its own detriment.[28] In doing so it has implicitly abandoned the idea that anything in the universe can be truly new. It is this dogma, not science itself, that stands in opposition to the essential biblical belief in God. In a materialist venue every scientific finding is just one more monotonous unearthing of what we already knew about the inner essence of things. Life, for example, is really "just chemistry." Mind is nothing more than matter parading itself under peculiar organic conditions. And the world deep down is really nothing more than a set of timeless physical routines *masquerading* as matter, life, mind, and spirit.[29]

Christianity, on the other hand, believes in an eternal freshness of being. Its God is one "who makes all things new" (Rev 23:5). Its expectation is that a new world is already being created. So Christian faith, though not irreconcilable with science, *is* irreconcilable with the modern materialist naturalism which logically rules out any such novelty. The really important disagreement, therefore, lies between naturalist physicalism, on the one hand, and a belief that the world can be made new, on the other.

SCIENCE, FREEDOM, AND THE FUTURE

Science is not the same thing as scientism, the belief that science alone can provide in principle an adequate understanding of everything. And, as I have already tried to make clear, science differs from scientific naturalism, the belief that nature is all there is. In fact, science from now on can become even more insightful if it casts off the antiquated materialist naturalism that holds it back from open-ended inquiry into the new. If it were to abandon the narrowly physicalist framework that has been its home for over three centuries, scientific exploration could be made compatible with a worldview that allows the world to manifest itself as truly unprecedented. In doing so it would also ensure that scientific inquiry will always have a future. This book, therefore, will attempt to express what some of the main discoveries of science can mean if we interpret them in the light of Christian expectation. It will be a *theology of nature* rather than a natural theology. Natural theology tries to show what nature, as scientifically known, can tell us about the existence of God. As I see it, however, a Christian theology of nature tries to express what the natural world means when we take it to be grounded in the reality of the God who in Christ and through the Spirit makes all things new.

---

28. Whitehead, *Science and the Modern World*, 51-59.

29. P. W. Atkins, *The 2nd Law: Energy, Chaos, and Form* (New York: Scientific American Books, 1994), esp. 200.

Christian faith, as it comes into encounter with science, need not present itself as terribly complicated. It is both simple and profound, but it does not have to appear convoluted. Even the doctrine of the Trinity does not have to be made needlessly arcane, although it will always remain mysterious. As a start we may express the substance of Christian faith in only three propositions. First, Christians believe in the reality of a transcendent mystery, the origin, ground, and destiny of the universe. We name this great mystery God. In Christian thought the all-encompassing origin, ground, and destiny of the universe is called the "Father," whom Jesus addressed intimately as *Abba*. Along with the faith of Israel, Christians understand this God as one who makes and keeps promises, breathes existence into all things, opens up the future, and makes all things new, even to the point of defeating death. As we shall see, science may seem at first to make the reality of a promising God questionable, but a Christian understanding of God provides human minds, including those of scientists, with limitless breathing room. God is the ground of freedom by virtue of being the world's future.

Second, Christianity instructs us that we should not think about God without first thinking about the man Jesus of Nazareth, the one who is called the Christ, the Messiah, the promised one who has become the foundation of all hope. The picture of Jesus given in the Christian tradition is that of a compassionately healing personality who, as the Son of God, was crucified but rose from the dead. Jesus is the very incarnation of God—the eternal divine Logos, the Word of God made flesh. As the risen Lord, the Son of God continues even now to open up the entirety of creation to a new future. Only in the context of a cosmic future centered on the risen Christ can we hope to enjoy the fullness of redemption and freedom.

Third, Christianity is about the work of the Spirit. According to St. John, early Christian writings, and the Nicene Creed, we understand the Spirit as the Giver of Life. The work of the Spirit is that of actualizing *emergence* in nature, of liberating life and consciousness from deterministic physical routine, and of sparking the urge toward freedom in human persons. The life of Jesus is one of being moved, even driven, by the Spirit of life and freedom. "It is for freedom," St. Paul says, "that Christ has set us free. Stand firm, then, and do not let yourselves be burdened again by a yoke of slavery" (Gal 5:1).

Each generation of Christians has to understand afresh what these words of St. Paul signify. Today we need to discern what they might mean in terms of scientific pictures of the world. The faith of Christians is a call to the fullness of freedom in the Spirit. But what could freedom possibly mean in an age of science? Science, after all, suggests that we are products of deterministic causal processes and hence that we are not really free after all. We are part of the physical universe where every event is the effect of invariant physical causes arising from the causal past. If we are part of law-bound nature, how could there be any room even for free choice, let alone for the exalted kind of liberty St. Paul is talking about?

This is one of many puzzles that science sets before theology today. In previous ages religions and philosophies often assumed that humans are not really a part of nature, so it was simpler then to think of ourselves as free. We were souls only temporarily encumbered by bodies. Today such a belief is incredible. Science has shown that we are completely continuous with natural processes even though our existence simultaneously extends toward what is not-yet. The physical universe gave birth to humanity no less than to other forms of life very gradually over the course of an immense amount of time. We are just as natural as bacteria, trees, and rodents. How then can we plausibly claim to be free without denying that we are also natural?

We need to face such questions squarely today. When I say "we" I mean not only Christians but other religious traditions as well. In this book I shall speak from a Christian perspective while simultaneously keeping in mind the implications of science for other traditions as well. I shall argue, in a manner that applies also to the wider story of religion on earth, that Christian theology must keep growing in the presence of scientific challenges, but without surrendering its hope for new creation. Today theology must become more attuned to biological evolution and the expanding universe, realities that for many sincere seekers have surpassed in magnitude and explanatory power our traditional ideas about God. A theology of nature, therefore, must show how theological reflection can provide a wide and generous ambience for the work of science. Christianity, of course, cannot introduce any new scientific information. It cannot determine whether this or that scientific idea is true or false. But a Christian theological setting can liberate science from belief systems— such as scientific naturalism—that make the world too small for both theology and open-minded scientific inquiry itself.[30]

THE PROMISE OF NATURE

In Christ the ultimate mystery that encompasses all created being is revealed as self-giving love and saving future. What then should we expect the universe to look like in light of the divine humility and promise that enfold it? My pro-

---

30. John Paul II, "Letter to the Reverend George V. Coyne, S.J., Director of the Vatican Observatory," *Origins* 18, 23 (November 17, 1988), 378: "Science can purify religion from error and superstition; religion can purify science from idolatry and false absolutes. Each can draw the other into a wider world, a world in which both can flourish."

posal is that a faith shaped by a Christian sense of God's self-limiting love, a love that lays open the future to new creation, should already have prepared our minds and hearts for the kind of universe that science is now spreading out in front of us.

Science has demonstrated over the past century and a half that the universe is a still-unfolding process and that it is unfathomably vaster and older than we had ever imagined before. The cosmos came into being long before the arrival of human history, Israel, and the church. Apparently God's creative vision for the world extends far beyond terrestrial precincts and ecclesiastical preoccupations. Nevertheless, a Christian theology of nature emanates from and tries to remain faithful to the teachings of the community of hope known as the church. Inspired by the "cloud of witnesses" (Heb 12:1) that has kept hope alive since Abraham, it wagers that the promissory perspective of biblical faith that enlivens people of faith is applicable also to cosmic reality in all of its enormous breadth and depth. "Thou hast made thy promise wide as the heavens," the psalmist exclaims (Ps 138:2, *New English Bible*). So, at least from the perspective of biblical faith, all those billions of years that preceded the emergence of Israel and Christianity were already seeded with promise.

With the eyes of hope one may still apprehend a vein of promise in this ambiguous cosmos. Genuine hope does not lead to escapist illusions but instead opens a space in which the scientific mind can breathe more freely than in the stagnant atmosphere of modern materialism. Hope will allow us to see that the world given to us by contemporary science has always possessed an *anticipatory* character. It has always been open to future surprise, though naturalistic pessimism has failed to notice. From the beginning the universe has extended itself toward the actualizing of new and unprecedented possibilities. It is still doing so, especially through one of its most recent evolutionary inventions, human consciousness. Through our own forays of hope, the universe now continues to seek out its future, a future whose ultimate depth we may call God.

After the phenomenon of mind had burst onto the terrestrial scene, the world's emergent straining toward the future took the form of religious aspiration everywhere. In the West it broke through in the hope we associate with Abraham, the prophets, and Jesus. Science itself, because of its orientation toward earlier and simpler lines of causation, knows nothing of any promise in nature, nor should we expect it to. Nevertheless, even though science cannot accurately predict the actual shape of the real novelty that will emerge in the future, this is no reason to assume that future cosmic happenings will somehow contradict the laws of physics, chemistry, and biology. The predictable habits of nature will go on functioning as before, but they will be taken up into an indefinite array of novel configurations.

THE MEANING OF MIRACLES

It is in terms of this promissory understanding of nature, it seems to me, that theology can speak most appropriately about miracles. Often the biggest obstacle to a scientifically educated person's acceptance of Christian faith is that it speaks of signs and wonders that seem to violate the inviolable laws of nature on which science depends for its own credibility. Later on we shall note, for example, that Albert Einstein rejected all forms of biblical theism because its devotees believe in a personal, responsive God who is said to be able to answer prayers and work miracles. To Einstein the existence of a supernatural agent who can intervene in the law-bound world and suspend its predictable operations is incompatible with science. Science has to assume, he insisted, that nature admits of no exceptions whatsoever. After all, what would be the point of scientists' articulating the unchanging laws of physics if nature could take off in unpredictable directions any time it—or God—was so inclined?

Theologians, I am convinced, must be sensitive to the fact that for scientifically educated people belief in miracles is a great obstacle to faith. We must be honest enough to ask whether Christian instruction, by insisting on a simplistic and literalist understanding of miracles, has tossed a false stumbling block onto the path of many who in other respects may be deeply attracted to Christianity. The same concern applies also to the matter of how to interpret the resurrection of Jesus from the dead. By giving people the impression that the resurrection and miracle stories are literally "violations" of nature,[31] have we perhaps distracted ourselves from their real meaning and at the same time made them appear unnecessary impediments to genuine faith?

In pulling together a response to these questions, an appropriate starting point might be to understand the resurrection of Jesus, and by analogy his own life and works of power (sometimes called miracles), as violations not of nature or science but of any worldview that makes deadness the most fundamental and "normal" state of being. As I shall argue later and at more length, the modern world has harbored, along with many other strains of thought, an "ontology of death." Both theologian Paul Tillich and philosopher Hans Jonas apply this formidable designation to the modern naturalistic assumption that everything alive came from, is explainable by, and is destined to return to, a state of absolute lifelessness.[32] It is not just the death of Jesus or of the total-

---

31. E.g., Norman L. Geisler, *Miracles and Modern Thought* (Grand Rapids: Zondervan, 1982), 12.

32. Hans Jonas, *The Phenomenon of Life* (New York: Harper & Row, 1966), 9; Tillich, *Systematic Theology*, 3:19.

ity of humans that the resurrection overturns. It also opposes any under-standing of the universe that gives explanatory finality to what is dead. Epis-temologically speaking, therefore, our minds need to be transposed into an anticipatory key in order to become attuned to the implications of Jesus' res-urrection as well as the meaning of miracle stories in the Bible.

An ontology of death maintains that the most probable, natural, and intel-ligible state of being is death, not life. A resurrection faith contradicts such a sense of reality, aware that the chronological priority of lifeless matter in nature's history, or its extensiveness in space, is not the same as being onto-logically foundational. Moreover, resisting the grounding assumptions of materialism does not entail opposition to science. Christian faith contradicts any worldview that forbids the breaking in of a *novum ultimum* ("what is ulti-mately new"), but this does not carry with it the implication that it opposes science. Whatever more specific interpretation we might give the New Testa-ment accounts of Jesus' deeds and his being raised from the dead, we miss their point if we fail to realize that they are thoroughly eschatological. It is the breaking in of a radically new future in the person of Jesus that the New Tes-tament authors are trying to communicate in their accounts of Jesus' life, words, works of power (*dynameis* in Greek), and resurrection from the dead. But this breaking in of the future does not mean the breaking down of nature, so science as such is not disturbed.

In order to hear properly the evangelists' or St. Paul's witness to the risen Lord, perhaps we need to have our thoughts and sensibilities transformed from the bottom up by the Bible's anticipatory worldview. However, it is just this worldview, a metaphysics of the future, that is the true stumbling block to Christian faith in an age of science. Requiring scientists to accept the res-urrection simply as a past event that violated nature and the laws of science only raises a needless barrier to embracing a life of Christian hope. It is both unnecessary and misleading to make scientifically educated people swallow the idea that the redemptive acts of God are fundamentally an interruption of the continuum of cause and effect in nature, or to ask them to believe that God has to suspend the laws of physics in order to answer our prayers.

However, by removing these obstacles, a theology of nature will not make Christianity easier to accept. It will require a larger leap than ever, but at least it is a leap that will not require the repudiation of the well-established results of empirical science. Something drastic, world-shaking, and soul-shattering will be required by embracing a resurrection faith, but intellectual integrity will not have to be compromised. After encountering the *real* challenge of Christianity it may even seem easier and more tempting, though certainly not as adventurous and exciting, to revert to the notion of miracles and resurrec-tion as though they were simply God's way of showing that the laws of nature can be broken. Christianity demands something much more consequential than such credulity. It demands the transformation of our whole understanding of the universe if we are to arrive at an appropriate understanding of God, nature, and ourselves. This transformation can occur, I believe, without our having to reject or edit anything we have learned from science. It is not science that is at stake in this process of conversion. What is at issue is the firmly entrenched ontology of death that underlies the naturalistic worldview. What will be threatened is the common assumption that we can find in the state of lifelessness and mindlessness a fundamental understanding of nature.

In brief, Christianity's real invitation is to let ourselves be grasped and shaken by the power of the future that is now and always dawning. The challenge to accept the news of Jesus' resurrection and his works of wonder is of a piece with the summons to believe that the entire universe is undergoing creative transformation. The truly important challenge of faith is to resist the always strong, but simplistic and enervating, inclination to view the world as resting on the dead physical past, and to learn instead to realize, as Teilhard has put it, that the world rests on the future as its sole foundation. The real challenge of Christian faith in an age of science is to realize the ontological primacy of life over the deadness that materialists take to be the normal, natural, and most intelligible state of being. I shall develop these reflections further in chapters 8 and 9.

SUGGESTIONS FOR FURTHER READING AND STUDY

Kelly, Anthony. *Eschatology and Hope.* Theology in Global Perspective. Maryknoll, N.Y.: Orbis Books, 2006.

King, Thomas M. *Teilhard's Mass: Approaches to "The Mass on the World."* New York: Paulist, 2005.

Moltmann, Jürgen. *Theology of Hope: On the Ground and the Implications of a Christian Eschatology.* Translated by James W. Leitch. New York: Harper & Row, 1967.

Peters, Ted. *God—The World's Future: Systematic Theology for a New Era.* 2nd ed. Minneapolis: Fortress, 2000.

2

Science and Mystery

CHRISTIAN FAITH ARISES from the hearing of God's "word." But this hearing presupposes a background of silence in which the word has forever been enshrouded. The hearing of God's word could not occur apart from our tacit awareness of a hidden domain of *mystery* out of which the word is spoken. Nor could God's word stir up our reverence and awe apart from at least a vague awareness of an inexhaustible depth that remains unspoken. Without a sense of mystery, faith is flat.

But hasn't science done away with mystery? Or, at the very least, has it not made mystery less imposing than before? Isn't science now bringing everything out into the clear light of day? And isn't the objective of science to "eliminate" mystery, as a famous Harvard scientist once put it?[1] Maybe "mystery" is equivalent to what has not yet been fully clarified by scientific method. So as human knowledge grows, perhaps the realm of mystery will shrink and eventually disappear altogether?

The suppression of a sense of mystery is a relatively recent occurrence. During the brief human period of the earth's long history most people in most places have felt an incomprehensible mystery enveloping their lives and the natural world. They have named and tamed it in different ways, but a premonition of mystery has kept their world from seeming dimensionless. An inkling of endless horizons beyond the immediate world has led shamans, prophets, mystics, and visionaries to undertake some of the most fascinating journeys in the history of exploration.

But what if these journeys lead nowhere and mystery is just another name for an endless void? This indeed is the belief of those who claim that nature is all there is and that science is the only reliable way to get to know it. On the other hand, our religious traditions have taken for granted the everlastingness of mystery. Indeed, their discourse has no meaningful reference apart from mystery's elusive permanence. So before tackling the more specific questions that science poses to Christian theology—issues relating to cosmology

---

1. B. F. Skinner, *Beyond Freedom and Dignity* (New York: Bantam Books, 1972), 54.

and creation, evolution and providence, the chemistry of life and the creativity of God, scientific law and human freedom—one must first decide whether mystery is real or perhaps instead a pretentious label for an unutterable blankness that surrounds us and our world.

If Christian faith is to have credibility in an age of science, mystery must mean more than mere hollowness. In all of its slippery silence it must impress us as a fullness of being rather than a vacuous abyss. Mystery has to be immune to any process of erosion in the face of science's advances. As science grows in insight, a sense of the mysterious must be allowed to intensify rather than shrink. If science in any way contracts our sense of mystery's endless reach, as many modern intellectuals have proposed, then science would be unalterably opposed to religion.

Moreover, the idea of *revelation* cannot become theologically intelligible unless its recipients first acquire a "pre-revelational" appreciation of mystery. Apart from such a foundational attentiveness to mystery, any divine word that comes to us will fail to grasp hold of us or enliven us. Without a prior sense of a transcendent dimension wherein God's word and revelation have been "hidden for ages past" (Eph 3:9), the actual self-disclosure of God that Christians believe to have occurred in Christ could not be delivered with force.

THE PERSISTENCE OF MYSTERY

It used to be that mystery was palpable to most people. It was both near and distant, a hidden depth and an intimate presence. It was the abyss out of which the physical world came to birth and into which all perishing events were everlastingly folded. Incomprehensible mystery transcended the world, but it also embraced and permeated everything. Mystery was the ultimate "whence" of the world, the ultimate "whither" of all that passes through time. It was out of this infinite dimension of depth that Christian faith always professed to have beheld the very face of God coming to greet and fulfill the creation, a face revealed in the man Jesus.[2] Without the backdrop of infinite mystery everything in Christian faith would have seemed shallow. No room would have existed for receiving the world of creation as a boundless gift and promise yet to be fulfilled.

However, especially since the beginning of the scientific revolution the world has come to seem less mysterious than before, at least to many educated people. It is uncertain just how far human consciousness even in the West has

---

2. John A. T. Robinson, *The Human Face of God* (Philadelphia: Westminster, 1973).

been secularized,[3] but scientific naturalists have declared that the space for faith has receded in direct proportion to the advance of scientific understanding. Faith is impossible without mystery, but mystery seems, at least to many, to have vanished. And where has it gone? Is it merely hiding so as not to be noticed? Or was it perhaps never there to begin with? Maybe mystery is a fictitious product of human ignorance, destined to wane as knowledge waxes. Perhaps nature is reducible to what science can dissect and control. In that case any present attention to mystery is a dalliance that only delays the march of science.

In view of such impressions, then, the first step in our inquiry into the relationship of science to Christian faith must be that of asking whether mystery still somehow impinges on human awareness, even after the physicists, geologists, biologists, and astronomers have allegedly demystified the world. If we cannot sustain or regain a powerful impression today that the world and our lives still dwell in endless mystery, Christian faith, along, of course, with all other religious traditions, will seem empty and illusory. A sense of the holy mystery we call God would then be exposed as having no inner substance and as referring to nothing more than an imagined construct rooted in worldweariness and the human desire to escape from nature. Christianity's claim that something of ultimate importance has been "revealed" in Christ would in that case be a vaporous dream rather than a volcanic eruption. Revelation will scarcely grab our attention or have any disclosive power if it cannot survive the naturalistic disenchantment of the world that has shadowed scientific modernity.

I will argue, therefore, that mystery abides and that we humans still exist within its grasp, even when we try to take flight from it. An orientation toward mystery is a structural trait of human existence, not an optional appendage peculiar to prescientific laggards. People are natively open not only to the world but also to a transcendent otherness, long before they have any actual conviction of being addressed by a revelatory word.[4] A "general" revelation of mystery is offered to all of us prior to our ever being encountered in history by the "special" revelation associated with Christ and the promises of God. Before hearing the word of God we have—all of us—already felt the presence of mystery, even when trying to deny or suppress it.

St. Paul (Rom 1:19) uses the Greek verb *phaneröö* ("to make manifest") when he says of humans in general that "what can be known about God is

---

3. Andrew M. Greeley, *Unsecular Man: The Persistence of Religion* (New York: Schocken Books, 1985).

4. Wolfhart Pannenberg, *What Is Man*, trans. Duane A. Priebe (Philadelphia: Fortress, 1970), 1-13.

plain to them, because God has shown it to them." And in speaking of the
special revelation that occurs in Christ, he employs the same verb: "But now,
apart from law, the righteousness of God *has been disclosed* . . . through faith in
Jesus Christ for all who believe" (Rom 3:21-22). Paul assumes that even apart
from Christians' experience of their risen Lord there is a general revelation of
God's gracious presence to all people and that they should already have
responded to its reality.[5] Correspondingly, St. Luke, in the Acts of the Apos-
tles, portrays Paul as having said to the Athenians, "What therefore you worship as unknown, this I proclaim to you" (Acts 17:23).[6] Similarly, in the
Prologue to the Gospel of John, the Word of God is said to enlighten *everyone* (John 1:9). And the consensus of many Christian writers from the beginning has been that even if we never hear anything explicit about Christ, our
existence and awareness have nonetheless already been touched by the mystery he makes manifest. Even apart from the experience of Christian revelation we are already being drawn toward the inexpressible depth of being that
Christians refer to as God.

And yet, as I have already noted, to countless sophisticated thinkers science
seems to have banished mystery from the world of human experience.[7] To
many of our contemporaries the intuition of mystery has faded, mostly
because science has made known so much that was previously unknown. Science excels in showing that what initially seems remarkable is really quite
unremarkable. It assumes that once we have understood any phenomenon by
specifying its physical causes and the unchanging natural laws it obeys, it is
no longer a mystery. So theology must ask where it is that mystery might *persistently* impinge on our lives even in a scientifically understood world. While
nature is becoming more subject to our cognitional control, is there any place
that we can expect mystery to be lurking in undiminished plenitude? Even in
the age of science, can we still locate a dimension of infinite depth, and therefore a resilient referent for our words about God?

MYSTERY AND PROBLEM

For the hard naturalist, our sense of mystery will eventually disappear as science continues to expand human knowledge of efficient and material causes.
As the highly respected physicist Heinz Pagels put it not too long before his

---

5. Schubert Ogden, *On Theology* (San Francisco: Harper & Row, 1986), 26-27.
6. Ibid.
7. E.g., E. O. Wilson, *Consilience: The Unity of Knowledge* (New York: Knopf, 1998); and Peter
W. Atkins, *Creation Revisited* (New York: W. H. Freeman, 1992), 11-17.

untimely death, "Now that astrophysicists understand the physics of the sun and the stars and the source of their power, they are no longer the mysteries they once were. People once worshipped the sun, awed by its power and beauty. In our culture we no longer worship the sun and see it as a divine presence as our ancestors did." Pagels acknowledges that many people "still involve their deepest feelings with the universe as a whole and regard its origin as mysterious." But as physics moves forward, he avows, "the existence of the universe will hold no more mystery for those who choose to understand it than the existence of the sun." So, "as knowledge of our universe matures, that ancient awestruck feeling of wonder at its size and duration seems inappropriate, a sensibility left over from an earlier age."[8]

However, is Pagels talking about mystery or about *problems*? With Gabriel Marcel I believe we need to make a clear distinction here.[9] A "problem" is a temporary gap in our attempts to understand and know. Accordingly, a problem is rightly expected to shrivel and eventually fade away as scientists work on it. "Mystery," on the other hand, is much more than a label for our present ignorance. It is infinitely more than a blank space to be filled in by scientific work or by any other intellectual and technological achievements. In fact mystery is able to loom larger and larger in our experience the clearer our scientific understanding becomes, or the more congenially physical patterns in nature yield to our technological control. Mystery resembles a horizon that keeps moving forward ahead of us into the unreachable distance as we work on and eventually solve the more manageable problems near at hand.

Unlike mystery, a problem can ultimately be solved and disposed of by the application of human ingenuity. Mystery, on the other hand, will forever resist any such resolution. Instead of growing smaller as scientists grow smarter, the sense of mystery can actually grow larger and deeper, at least to those who allow themselves to be carried away by it. Unlike a problem, mystery cannot be contained within intellectually clear boundaries. Instead, it spills out past all our efforts at intellectual control. Problems can be disposed of. Mystery, on the other hand, wildly resists all efforts to bottle and cap it.

But, again, where do we actually find this wanton persistence of mystery in our experience and understanding? I shall address this question and give illustrations below, but first I want to make it clear that there are at least some prominent scientists who claim to be quite open to mystery. Perhaps the most notable example is Albert Einstein. For this great physicist mystery signifies a dimension of the universe that will forever remain incomprehensible to, and

---

8. Heinz Pagels, *Perfect Symmetry* (New York: Bantam Books, 1985), xv.
9. Gabriel Marcel, *Being and Having* (Westminster: Dacre Press, 1949), 117.

undiminished by, science. Mystery, he insists, will always be around, and so scientists will never be able to reach the end of their exciting journey of discovery. "The most beautiful experience we can have is of the mysterious," Einstein says. And "whoever does not know it and can no longer wonder, no longer marvel, is as good as dead." "It is this knowledge and this emotion," he continues "that constitute true religiosity."[10]

Even though Einstein rejects the idea of a personal God and denies the possibility of any special revelations, he still considers himself to be religious. For him, religion is the grateful cultivation of an awareness that the world is encompassed by an irremovable mystery. And the best evidence for the existence of mystery is that the universe is intelligible at all. Science itself can never grasp or understand why this is so, but can only take it as a given. It is especially on the question of why the universe makes sense at all that, for Einstein, scientific thought hits an unsurpassable barrier. Mystery remains—even after science.

LIMIT EXPERIENCES

Where, though, does mystery show up in the lives of humans less extraordinary than Einstein? One response is that we feel mystery most abruptly in the *limit experiences* that every person encounters sooner or later.[11] A limit experience, broadly speaking, is any occasion in which we lose the sense of being in control. This may be a moment of great joy, tragedy, or spiritual uncertainty. A limit experience may be an event in which fate, death, guilt, or doubt about the meaning of our lives threatens to overwhelm us. Most of us have felt, at one time or another, the blunt incursion of something unmanageable in our existence. We brush up against it whenever we cannot find clear and certain answers to questions about where we came from, what our destiny is, and what we should be doing with our lives. Moments of terror, guilt, and doubt, but also occasions of great joy, expose us in exceptional ways to limits that are always present but not always palpable. Our limit experiences give rise to "ultimate" questions. More mundane questions take up most of our lives, but at times earthquakes occur that seem to open up an abyss beneath us. These happen, for example, when we experience personal failure, fall seriously ill, or

10. Albert Einstein, *Ideas and Opinions* (New York: Modern Library, 1994), 11.

11. David Tracy, *Blessed Rage for Order: The New Pluralism in Theology* (New York: Seabury, 1975), 91-118; Tracy relies here in part on Stephen Toulmin, *An Examination of the Place of Reason in Ethics* (Cambridge: Cambridge University Press, 1970), 202-21.

are beset with grief at the death of someone we love. Such upheavals expose us to an abyss from which we instinctively recoil, but which can also lead us to a dimension of depth where hope conquers fear and sadness.

In the experience of limits we are invited to decide: Should we understand mystery as an infinite fullness, or as a fathomless void? Should we greet it as an absolute emptiness that makes all effort ultimately worthless? Or should we trust it, let it transform our lives and deepen our understanding of ourselves and the universe? Can we permit ourselves adventurously to be grasped by mystery, or should we instead cling to the banal security of the already mastered world?

Paul Tillich refers to mystery as depth, and in this depth, he says, there lies the prospect of unsurpassed joy.[12] Religions also generally teach that in the depths of a mystery too abysmal for minds to comprehend, tragedy can be transformed into triumph. The earliest Christians, for example, journeyed from despondency immediately after Jesus' crucifixion to a profound experience of his risen presence and the outpouring of his Spirit. Mystery is an abyss, but it is also ground. It repels but also attracts, at times irresistibly.[13]

Perhaps it is a vague, anticipatory grasp of the endlessly fascinating and fulfilling horizon of mystery that allows us to apprehend, by way of contrast, the ordinariness of our everyday worlds. It is only because of the structural openness of our being to what is "more" that we humans can experience the monotony of what is "less." Other species of sentient beings apparently never have the experience of an intolerable emptiness. For humans, however, the capacity to feel tedium in our lives is a consequence of our already having an inkling of the wide-open world of mystery hovering at the margins of the mundane, threatening—or promising—to renew and enliven us.

It is also our native anticipation of endless mystery that inspires us to imagine alternative worlds, whether in the form of fairy tales, utopias, or eschatologies. So, even in its crudest forms, our expectation of "more" is not entirely reducible to mere illusion. Maybe our imaginative leaps are not always merely childish wishing, even though there is probably a touch of naiveté in most of our visions of better worlds. But, at bottom, our visionary impetuosity may be the consequence of the fact that at some level of our being and awareness we have already been grasped by an infinite mystery. This is why our hearts are so restless, as St. Augustine observed. Mystery is not an absolute void, although it may seem to be so at first, but an elusive *plenitude* of being

---

12. Paul Tillich, *The Shaking of the Foundations* (1948; New York: Charles Scribner's Sons, 1996), 63.

13. Ibid., 52–63.

that we cannot grasp because it has already grasped us. What Christian theology calls "revelation" is the medium through which we are touched by and reconciled with mystery.

Much of the risk, adventure, and seriousness of human life consists in the fact that each of us must make a decision as to whether mystery is a bottomless void or a fullness too large, generous, and resourceful to comprehend. Is mystery an invitation to despair, or instead to hope? Religions have sought to help us make a decision about the character of mystery. They have looked for a *face*, a personality—and sometimes many personalities—emanating from the impenetrable mist that encircles us and our universe. At times they have fought bitterly with one another about which face will be focal, or about whether any face is appropriate at all. Religions have generally felt ambiguous about their images of mystery, sometimes clinging to them as idols, sometimes casting them out in order to be free of enslavement to them. No images, after all, can adequately reveal the silent depths of mystery, but symbols are essential if religions are to say anything about mystery at all. Consequently, religious life moves back and forth, sometimes tumultuously, between the extremes of idolatry and iconoclasm, certitude and doubt.

LIMIT QUESTIONS

Mystery also shows up in what the philosopher Stephen Toulmin calls "limit-questions."[14] These are questions that arise, for example, at the boundary, or limits, of scientific inquiry. Mystery lurks behind these limits. Most of the time when researchers are involved in the scientific quest for truth they are not reflectively attending to mystery. To do so would be a crippling distraction. Nevertheless, mystery peers in from the margins of their explorations, and it shows up dramatically, and all of a sudden, when scientists find themselves asking such questions as why they should be doing science at all. A sense of mystery is not an explicit part of the awareness of scientists while they are actively working on specific problems. Nevertheless, while driving home from work one day, the thoughtful scientist might suddenly ask: Why am I bothering to do science at all? What is the meaning of my work? Is it really worthwhile spending my days looking for truth? What is the point of it all?

These are limit questions, quite different from the solvable problems that occur *within* science. They arise only at the edge of scientific investigation,

---

14. Toulmin, *Examination of the Place of Reason*, 202-21.

and, unlike problems, they do not admit of any "solution." Limit questions never go away, since it is the abiding reality of incomprehensible mystery that invites a person to ask them at all. It is to these questions that religions properly turn their attention.[15] Unnecessary confusion arises, therefore, whenever people refer to sacred texts as though their purpose includes addressing what in fact are scientific questions, such as whether Pluto is a planet or whether string theory can add anything significant to our understanding of the universe. Religion and theology are best understood as responses to limit questions, not as solutions to particular problems that science can solve by itself.

This is not unlike a principle that Galileo articulated as long ago as the seventeenth century and which St. Augustine had formulated centuries earlier. In his remarkable "Letter to the Grand Duchess Christina," Galileo pointed out that some of his ecclesiastical opponents had wrongly assumed that the biblical authors intended to deliver accurate propositions about the natural world along with their religious message. The problem with this assumption, he pointed out, is that if the biblical impressions about nature are sometimes shown to be wrong, as Copernicus's ideas and his own discoveries implied, there will be nothing to prevent people from being suspicious about the truth of the Bible's religious instructions as well. For support Galileo referred to St. Augustine's *De Genesi ad Literam,* in which the great theologian had wisely stated that in reading Genesis Christians should not get hung up on questions of its astronomical accuracy. In their conversations with unbelievers, Augustine had argued, Christians should not search Holy Scripture for answers to such questions as "whether heaven, like a sphere, surrounds the earth on all sides as a mass balanced in the center of the universe, or whether like a dish it merely covers and overcasts the earth." Doing so will only lead unbelievers to suspect the truth of the biblical writings "when they teach, relate, and deliver more profitable matters."[16]

In the same vein Galileo wrote to the Grand Duchess that the Holy Spirit "did not intend to teach us whether heaven moves or stands still" since we can find that out on our own. So,

> if the Holy Spirit has purposely neglected to teach us propositions of this sort . . . how can anyone affirm that it is obligatory to take sides on them, and that one belief is required by faith, while the other side is erroneous? Can an opinion be heretical and yet have no concern with the salvation of

---

15. See Tracy, *Blessed Rage for Order,* 94-109; also Schubert Ogden, *The Reality of God and Other Essays* (New York: Harper & Row, 1977), 31.

16. *Discoveries and Opinions of Galileo,* trans. with introduction and notes by Stillman Drake (New York: Anchor Books, 1957), 186.

souls? Can the Holy Ghost be asserted not to have intended teaching us something that does concern our salvation? I would say here something that was heard from an ecclesiastic of the most eminent degree [apparently Cardinal Baronius (1538-1607)]: "That the intention of the Holy Ghost is to teach us how one goes to heaven, not how heaven goes."[17]

"Hence," Galileo concludes, "I should think it would be the part of prudence not to permit anyone to usurp scriptural texts and force them in some way to maintain any physical conclusions to be true, when at some future time the senses and demonstrative or necessary reasons may show the contrary."[18]

And yet, even today scientific naturalists no less frequently than biblical literalists still interpret religious doctrines and Scriptures as though it is part of their intention to solve scientific *problems*. This misunderstanding pertains especially to contemporary controversies about Christianity and evolution. For example, the evolutionist Gary Cziko claims that it is natural selection *rather than* divine providence that accounts for why we have ears, eyes, and minds. He assumes that for religious people the biblical notion of divine providence has functioned all along as the solution to a scientific kind of question, and that any appeal to providence has become obsolete now that Darwinian explanations are available.[19] Likewise, some religious opponents of evolutionary biology try to force the notion of "intelligent design" into biology classrooms. Heedless of Galileo's advice, they are taking what is in fact a response to a limit question (why is there any order at all?) and making it serve as the answer to a specific set of problems that science can address without explicit reference to God. Good science as well as sound theology both rightly object to this ruse.

In thinking about Christianity and science, therefore, it is of utmost importance that believers understand the sources of their faith not as a response to scientific problems, but instead as a way of addressing limit questions. Divine revelation is trivialized whenever its content is wedged into explanatory slots reserved for scientific understanding. Although Christian faith provides a vision of reality that encourages one to seek scientific understanding, the content of its religious viewpoint is no part of science as such. The articulation of scientific truth is never the business of theology. Of course, what science finds out about the universe is not a matter of indifference to theology. Evolutionary accounts of life, for example, cannot help but change

---

17. Ibid., 185-86.
18. Ibid., 187.
19. Gary Cziko, *Without Miracles: Universal Selection Theory and the Second Darwinian Revolution* (Cambridge, Mass.: MIT Press, 1995).

how one thinks about the world and its Creator. And knowing about the Big Bang and an expanding universe can give new depth to the traditional idea of the cosmic Christ. Inevitably science does have an impact on religious thought. But the tasks of theology and science remain distinct.

To summarize, then, it is especially at the limits of human experience and scientific problem solving that we encounter the incomprehensible mystery that gives rise to religious striving and symbolizing. It is at these limits that we may find ourselves suddenly asking unsolvable religious questions. We cannot help asking them since mystery, in a "general" way, has already "revealed" itself to us in our encounter with limits.

MYSTERY AND SPECIAL REVELATION

Our encounter with the "original" revelation of mystery leads us to look for a "special" revelation by which to experience more humanly and more palpably what kind of mystery it is that *always* touches us elusively.[20] Much of our life involves an ambivalence about mystery, so the search for a "revelation" that might make things less general and more concrete is essential. Humankind's religious quests instinctively look for symbols that will bring into focus the original revelation of mystery that lies at the limits of our lives and awareness. "At the base of every religion, as its origin and principle," theologian Schubert Ogden writes, "is some particular occasion of insight, or reflective grasp through concept and symbol, of the mystery manifested in original revelation."[21] To Christians this special disclosure of mystery occurs decisively in Jesus.

The term "revelation" comes from the Latin *revelare*, "to remove a veil." Since around the middle of the first millennium B.C.E., some religious and philosophical traditions have taught that a veil of illusions ordinarily obscures from us the true depth and character of mystery. Consequently, these same traditions have laid out various paths by which the everyday clouding over of consciousness can be broken through. These paths may be thought of, theologically speaking, as in search of a *special* revelation that may help dispel mystery's ambiguity.

---

20. Ogden, *On Theology*, 41: "*what* Christian revelation reveals to us is nothing new, since such truths as it makes explicit must already be known to us implicitly in every moment of our existence. But *that* this revelation occurs does reveal something new to us in that, as itself event, it is the occurrence in our history of the transcendent event of God's love" (p. 43).

21. Ibid., 40.

By "special revelation" I mean a concrete symbolic disclosure, culturally and historically conditioned, of the universal and eternal mystery of God. Christianity is a religion that finds the divine mystery irresistibly and powerfully presented in the person of Jesus of Nazareth.[22] The appearance in history of this man has brought about nothing less than a revolution in our understanding of holy mystery. Christians in fact are instructed not to think about mystery without thinking also about Jesus. Even so, throughout the history of Christianity there have been many different ways of portraying this man and his mission. The various books in the New Testament themselves picture Jesus in different ways: as an infant in the manger, a teacher of wisdom, the Word of God, a wonder-worker, a healer, a prophet, a crucified victim, the Son of Man, a risen savior victorious over death, the one who is coming, and a cosmic figure gathering the whole universe to himself. Christians believe that in Jesus the Christ the fullness of infinite mystery has burst decisively into our midst and welcomed us warmly into its embrace. Christian creeds claim, first, that the world is grounded in the unspeakable mystery of God the *Father*, the Creator of the heavens and the earth; second, that this silent mystery paradoxically gives itself to the world unreservedly in the person of Jesus who is called the *Son* of God; and, third, that by the work of the *Holy Spirit* the infinite mystery lovingly gathers the world into itself, liberates life, renews the creation, and saves all that is lost.

It is only of one God that Christian faith speaks, but from the very beginning, Christian prayer and reflection have mysteriously sought to avoid simplistic representations of God. To protect the polyvalence of their experience of the divine, the early Christians developed—quite unsystematically—a trinitarian way of speaking about God. Later attempts by theology to systematize with intellectual clarity the mystery of the Trinity have never been fully successful even to this day. Trinitarian language is essential to the way in which Christians unfold the richness and depth of the mystery that has been revealed in Jesus, but theology's grasp of this mystery remains tenuous—and forgivably so.

We humans have an insatiable need to make sense of the mystery that underlies personal existence, human history, and the universe itself. Hence, as H. Richard Niebuhr has noted, we are looking for a revelatory "image" that will help us understand and name the mystery of being.[23] Christians believe that we are graced with such an image in the picture of God conveyed by the

---

22. See Paul Tillich, *Systematic Theology*, 3 vols. (Chicago: University of Chicago Press, 1963), 1:147-50.

23. H. Richard Niebuhr, *The Meaning of Revelation* (New York: Macmillan, 1960), 80.

person and work of Jesus. The doctrine of the Trinity has emerged as an essential conceptual matrix of Christian attempts to understand how God can be said to be fully manifested in Jesus. Consequently, it would be misleading to understand the Trinity apart from its being anchored in the primordial Christian belief that in Christ God has "bent down"[24] to embrace the world and, in doing so, has opened up for it a new future.

GOD AND THE UNIVERSE

Christian tradition proposes, paradoxically, that the encompassing mystery of God has been made manifest in the life of a unique human being at a particular time in human history. But in an age of science, how are we to connect the triune mystery, the strange ideas about God that began to emerge during the earliest Christian experiences of the risen Christ, to our new understanding of the universe? We are far from finished with this most significant theological project. There is material in the last century and a half of science to spark a whole new revolution in theology, but so far the ember has barely caught fire. Apart from a few great Christian thinkers whose thoughts I shall be interpreting and extending in these pages, theology has been reluctant to take advantage of the most exciting discoveries of science.

I noted earlier that some scientists see their work as an attempt to eliminate mystery. Ironically, however, the deeper science digs, the more impressive is the extent of the mystery it uncovers. But theology has yet to take fully into account what science has churned up from the depths of nature. This is true not only in the case of Copernicus and Galileo, but also Darwin, Einstein, Hubble, Heisenberg, and Hawking. Theology for the most part is still fixated on other issues—worthy ones of course—but at the expense of failing to generate a spiritually electrifying theology of nature. Were it to look more deliberately at the new ideas about life and the universe, Christian theology might find that its revelatory image of God is able not only to accommodate but also to render even more intelligible the information that keeps coming in from the work of scientists. Christianity, if we but highlight its most distinctive features, can be a most illuminating context for understanding what is going on in the universe and not just human history.

What then does Christian revelation tell us that would help us better understand the universe as it is now also being revealed by science? Obviously

---

24. See Ilia Delio's lovely and persuasive book *The Humility of God* (Cincinnati: St. Anthony Messenger Press, 2005).

theology has nothing to offer in the way of scientific information and problem solving. In the spirit of Augustine and Galileo it is essential to remind
ourselves always that theology cannot add anything to the burgeoning mound
of scientific facts or theories. Nonetheless, I will argue that the revelatory
"image" of God's descent and futurity can now be extended beyond its application to history and human persons so as to illuminate also the natural world
that is the wider context of both. In each subsequent chapter I shall spell out
this proposal more specifically.

THE PROBLEM OF SPECIAL REVELATION

Before proceeding, however, I must comment further on the fact that according to scientific naturalists science has not only debunked the pre-revelational
sense of mystery but also the prospect of any special revelation such as Christianity claims. There is considerable doubt among the scientifically educated
these days about whether any special revelation either has actually happened
or can take place at all. Would not such an event tear an arbitrary hole in the
seamless fabric of nature as science has come to know it? As Einstein and
other scientific naturalists insist, any special revelation of God would rupture
the tissue of natural occurrence and bring an end to scientific trust in nature's
remorseless consistency.

Throughout most of human history interruptive divine revelations seemed
commonplace. Even the dreams of ordinary people seemed to let in the reality of other worlds. But modern science apparently has no room for such
exceptional events. It is true that popular culture today allows for appearances
from the "beyond," but to most educated people the possibility of miracles
and special revelations has become questionable. The claim that profane or
secular experience can be perforated by a word of God that breaks into our
midst to judge our sinfulness and show us the path to salvation is now unbelievable to many. Religions and revelations appear to be little more than our
own creations. Consequently, the Christian instruction that we should allow
our lives to be transformed by a special revelation often seems bizarre.

Paul Davies, an acquaintance of mine and a scientist who is not at all
unfriendly toward religions, illustrates my point. "Science," he has written, "is
based on careful observation and experiment" whereas "religion is founded on
revelation and received wisdom." Moreover, revelation "is liable to be wrong,
and even if it is right other people require a good reason to share the recipients' belief."[25] Davies's position will strike most theologians as a caricature,

---

25. Paul Davies, *God and the New Physics* (New York: Simon & Schuster, 1983), 6.

but it illustrates how problematic the concept of revelation is among the scientifically educated. Interestingly, even to some theologians the idea of revelation, in the words of Protestant theologian Stanley Hauerwas, "creates more trouble than it is worth."[26] Ronald Thiemann, who disagrees with Hauerwas, nevertheless adds that "most discussions of revelation have created complex conceptual and epistemological tangles that are difficult to understand and nearly impossible to unravel."[27]

The majority of Christian theologians, on the other hand, are still willing to affirm the reality of revelation. Catholics among them would cite one of the most important documents of the Second Vatican Council, *Dei Verbum,* in which the topic of revelation is central. Nevertheless, theologians need to face frankly the disaffection with revelation in the contemporary intellectual world, since for many people today the idea seems incompatible with science.[28] We may begin by presenting the meaning and content of revelation in such a way as not to make them seem incompatible with science. The following chapter will attempt to show how this is possible.

SUGGESTIONS FOR FURTHER READING AND STUDY

Einstein, Albert. *Ideas and Opinions.* New York: Modern Library, 1994.

Haught, John. *What Is God? How to Think about the Divine.* Mahwah, N.J., and New York: Paulist, 1986.

Murchie, Guy. *The Seven Mysteries of Life: An Exploration in Science and Philosophy.* Boston: Houghton Mifflin, 1978.

Pannenberg, Wolfhart. *What Is Man?* Translated by Duane A. Priebe. Philadelphia: Fortress, 1970.

---

26. As quoted by Ronald Thiemann, *Revelation and Theology* (Notre Dame: University of Notre Dame Press, 1985), 1.

27. Ibid.

28. I shall take up the question of revelation and truth in chapter 10.

3

Science and Revelation

CHRISTIANITY IS A RELIGION whose teachings are said to come from a special *revelation*. But what is revelation? Is not revelation a set of events that interrupts and contradicts the natural manner in which things take place? Is Einstein perhaps correct in his assertion that revelation is incompatible with science? Revelation, like miracles, seems to breach the tightly closed continuum of causes and effects that make up the natural world, at least as science sees it. Consequently, many scientists and philosophers—especially those I have been referring to as scientific naturalists—find Christianity, as well as other revealed religions, unbelievable.

This book, as the reader will already have noted, is a theological conversation not only with believers, scientists, and other inquisitive people, but also with scientific naturalists, the main contemporary representatives of what Friedrich Schleiermacher called the cultured despisers of religion.[1] As I have already pointed out, it is not science as such but the modern belief system of scientific naturalism that rejects the possibility of any special revelation. And so it is important that contemporary theological reflections on nature remain aware of the naturalistic belief system and its claims that theology has no business commenting on nature at all. According to scientific naturalists, reliable understanding must be publicly accessible and subject to empirical testing. Since revelation does not submit to such criteria, naturalists dismiss Christianity as illusory.

Christians understand their faith as a response to the divine mystery that presents itself in the person, life, words, actions, death, and resurrection of Jesus. But I think we can say that this revelation is an eruption, not an interruption, of nature. Accordingly, revelation is not a violation of nature's inviolable routines, but an expression in symbolic terms of a deep and momentous drama always going on in the depths of the universe and human history. The reality and power of this drama are inaccessible to science and can only become part of the worldview of people of faith who have allowed themselves

---

1. Friedrich Schleiermacher, *On Religion: Speeches to Its Cultured Despisers* (New York: Harper & Row, 1958).

to be swept up into it. Revelation, like science, is about what is *really* going on in the universe, but it discloses to faith a dimension of reality that necessarily goes unnoticed by scientific inquiry. It does not contradict science, but it does call for our acknowledging the limitations of science.

Theology must respect the integrity and autonomy of science, but at the same time it may question whether science alone can capture everything that is going on in the universe. A theology of nature does not deny that science can put the human mind in touch with nature or that science can reveal previously unknown things about the universe. Yet, without taking anything away from science, it proposes that there are levels of depth in nature that science simply cannot reach. A theology based on revelation does not compete with science or conflict with it in any way. It is complementary to it, in the sense that it contributes something to the larger picture of reality that science cannot. A theology of nature construes the universe in such a way as to support the work of science while at the same time refusing to confuse science with scientism and scientific naturalism.

Like science, revelation implies that there is always much more to the world than what *seems* to be the case.[2] Good scientists are willing to abandon or revise their hypotheses and theories whenever they sense that there is deeper intelligibility beneath their all too simple models and mathematical calculations. Hearers of a divine revelatory word, for their part, are also obliged at times to reach for fresh symbols, or perhaps lapse into complete silence, in the presence of the incomprehensible mystery that grounds, sustains, and fulfills the universe.

The God of Christian faith is encountered in and through the observable world. So revelation unveils the divine mystery by way of symbols derived from our experience of nature and social existence. Revelation occurs not only in words but also in sacraments derived from nature. That is, natural phenomena, not just events in human history, participate in and thus point us toward the mystery of God. Water, light, food, soil, fertility, life, and human personality are indispensable to the experience of revelation. God is known not apart from nature but in and through it. By virtue of the incarnation the entire drama of nature unfolding across billions of years is also the revelation of God.

THE GIFT OF AN IMAGE

In the Christian context, however, revelation is fundamentally the gift of the infinite mystery of God's own being *to* the finite world. The primary sacra-

---

2. Huston Smith, *Forgotten Truth: The Primordial Tradition* (New York: Harper & Row, 1976), 97.

ment or symbol of God's self-gift is the person of Christ,[3] but scientific awareness, today more than ever, allows us to assume that the entire universe, by virtue of the incarnation, is tied inseparably into the revelation of God in Christ. Contemporary astrophysics compels us to acknowledge that the emergence of human life with its capacity for thought, morality, hope, and worship is seamlessly connected to the birth and development of an entire universe. Over the last half-century we have learned more and more about the cosmic conditions—starting at the first instant of cosmogenesis—that had to be in place if life and thought were ever to exist at all. The cosmos and consciousness can no longer be dualistically split off from each other. Theologically, the new scientific awareness means that the appearance in our midst of the person of Christ, therefore, is not just a historical but also a terrestrial and cosmic event. From now on when we tell the story of Jesus we need to include not only its biblical setting but its natural prelude as well.

Theology until recently has not had the opportunity to tie all the ages of the cosmos so tightly to God's revelation in Christ. Now, however, science permits Christians to connect their faith to the wider world of human and cosmic history, as expressed for example in these words of Pierre Teilhard de Chardin:

> The prodigious expanses of time which preceded the first Christmas were not empty of Christ: they were imbued with the influx of his power. It was the ferment of his conception that stirred up the cosmic masses and directed the initial developments of the biosphere. It was the travail preceding his birth that accelerated the development of instinct and the birth of thought upon the earth. Let us have done with the stupidity which makes a stumbling-block of the endless eras of expectancy imposed on us by the Messiah; the fearful, anonymous labors of primitive man, the beauty fashioned through its age-long history by ancient Egypt, the anxious expectancies of Israel, the patient distilling of the attar of oriental mysticism, the endless refining of wisdom by the Greeks: all these were needed before the Flower could blossom on the rod of Jesse and of all humanity. All these preparatory processes were cosmically and biologically necessary that Christ might set foot upon our human stage. And all this labor was set in motion by the active, creative awakening of his soul inasmuch as that human soul had been chosen to breathe life into the universe. When Christ first appeared before men in the arms of Mary he had already stirred up the world.[4]

---

3. Edward Schillebeeckx, *Christ, the Sacrament of Encounter with God,* trans. Paul Barrett and N. D. Smith (New York: Sheed & Ward, 1965).

4. Pierre Teilhard de Chardin, *Hymn of the Universe,* trans. Gerald Vann, O.P. (New York: Harper Colophon, 1969), 76-77.

Revelation is much more, therefore, than the *locutio Dei*, the "speech of God." It is more than what St. Augustine called the divine "illumination" of our souls. Revelation's first meaning is not the passing on to us of propositional truths from God. It is not simply "the communication of those truths which are necessary and profitable for human salvation . . . in the form of ideas."[5] Nor is revelation reducible to "direct discourse and instruction on the part of God." It means much more than "an act by which God exhibits to the created mind his judgments in their formal expression, in internal or external words."[6] Revelation, before everything else, is the gift of God's own being and selfhood to and through the entire *universe*.

Theologian Karl Rahner writes that the "primary phenomenon given by faith is precisely the self-emptying of God."[7] A theology of nature, therefore, will ask whether this shocking theological proposal can help people of faith make sense of what science is now finding out about the physical universe. Contemporary theology, both Catholic and Protestant, increasingly interprets revelation to mean the gift of God's own self to the world. This is an idea that flickered feebly even at the First Vatican Council, which says in its Constitution on Faith, "It has pleased God to reveal himself and the eternal decrees of his will to the human race" (chap. 2),[8] but until not too long ago Christian theology usually featured an excessively intellectualized and propositional understanding of revelation. Today, as the result of a closer reading of the Bible and other traditional theological sources—and thanks especially to the Second Vatican Council—theology has been moving toward the view that the actual content of revelation is the infinite mystery of God's own being.

This content, however, comes to faith in the first place not in theological formulas but in startling images. As theologian H. Richard Niebuhr puts it, revelation is "the gift of an image." Revelation is "that special occasion which provides us with an image by means of which all occasions of personal and common life become intelligible." A truly revelatory image must be able to render meaningful what might otherwise seem meaningless. Niebuhr goes on to say that the revelatory image offers a "pattern of dramatic unity . . . with the aid of which the heart can understand what has happened, is happening and will happen to selves and their community."[9] A theology of nature will add,

---

5. From P. Schanz's *Apologie des Christentums* (1905), quoted by Werner Bulst, *Revelation*, trans. Bruce Vawter (New York: Sheed & Ward, 1965), 18.

6. B. Goebel, *Katholische Apologetik* (1930), as quoted by Bulst, *Revelation*, 18.

7. Karl Rahner, *Foundations of Christian Faith*, trans. William V. Dych (New York: Crossroad, 1978), 222.

8. See Josef Neuner and Jacques Dupuis, eds., *The Christian Faith*, 7th ed. (Staten Island, N.Y.: Alba House, 2001), 43; Bulst, *Revelation*, 23.

9. H. Richard Niebuhr, *The Meaning of Revelation* (New York: Macmillan, 1960), 80.

however, that a truly revelatory imagery illuminates not only human history and social existence but also the entire universe.

Perhaps the reception of revelation is analogous to what happens in science when a new insight suddenly flashes into our awareness, shedding light on previously unsolved problems in a most surprising way. For example, when Copernicus and Galileo redrafted the cosmological map, construing the heavens as heliocentric rather than geocentric, numerous difficulties associated with the older Ptolemaic system suddenly went away, and more intelligible models of the heavens abruptly took their place. Imaginative breakthroughs in science have the effect of bringing to light previously hidden aspects of nature. Theories associated with Newton, Darwin, and Einstein, along with the more recent ideas in physics and geology, have all brought a surprisingly fresh coherence to our understanding of nature. And the new models have had the effect of leading to further fruitful research. Today scientists are looking for an elegant formula that will tie together the four main physical forces in nature.[10] When and if such a theory comes along we can safely predict that it will lead not to the end of science, as some fear. Rather it will come as a gift that opens up surprising new areas for ongoing research and discovery.

In order to be of interest to us, revelation must at the very least have a similarly startling, illuminating, and fertile effect. Its content must shake our understanding of reality, including our understanding of the natural world, but in such a way as to make it more, not less, intelligible. Revelation will not compete with scientific understanding, but in order to qualify as a momentous event of disclosure it must help us make sense not only of our personal lives and human history but also of the general features of the cosmos itself.

Revelation has no answer to specific scientific questions such as how life evolves, or what the mechanisms of evolutionary change might be. Revelation, as I noted in the previous chapter, responds to limit questions, not unsolved scientific problems. For example, when we ask why we should bother to do science at all, revelation may at least be able to help us understand in more depth why truth is worth seeking. Perhaps it may also be able to shed light on why nature and life are subject to evolution at all, why intelligent life appeared in the universe, why the universe is such as to allow new things to happen, and why it is such an exquisite blend of accidents, laws, and deep time as to unfold in the narrative way that allows it to be the bearer of meaning.

The "gift of an image" that Niebuhr associates with revelation will not provide any information that adds to the mound of scientific ideas, nor will it compete with science in any way. Yet revelation may still provide an enliven-

---

10. Stephen Hawking, *A Brief History of Time* (New York: Bantam Books, 1988), 155-69.

ing sense of the *meaning* of the universe that science is now setting before us. What would be the point of our making so much of "revelation," after all, unless it has the power to make things, including the discoveries of science, even more intelligible than before?[11] Once again, revelation must be able to shed new light not only on human selfhood and human history but also on the universe.

CHRISTIANITY'S REVELATORY IMAGE

But where, more specifically, can we encounter a revelatory image that would justify such an ambitious expectation? A Christian theology of nature may find its starting point in the image of a self-giving, promising God as made manifest in Jesus of Nazareth. On the one hand, Christian faith presents us in Jesus the image of an infinitely loving God who bends humbly toward the creation in order to relate to it with unsurpassable, incarnate intimacy. On the other hand, the Scriptures canonized by Christian tradition, in narratives ranging from those about Abraham to those about Jesus' resurrection, give us the picture of a God who makes promises and in doing so opens up the whole of created being, not just human history, to an always new future.

Reflecting on the natural world as science understands it today, we may find that the two related themes of the "descent of God" and the "futurity of God" are most instructive. The revelatory disclosure in Christ of God's humble self-emptying may help us understand aspects of nature—not least its very creation, its processive character, and the evolutionary way in which life develops—that would otherwise remain ultimately unintelligible, no matter how far science goes in disclosing the details of natural occurrences. At the same time, the biblical image of a God who makes promises and remains faithful to them can help our theology of nature make new sense of the whole suite of scientific discoveries associated with the idea of *emergence*.

Lighting up our reflections in this and subsequent chapters will be the twin beacons of (a) God's humble self-abnegation, an eternal gesture of descent that somehow allows the universe to exist and evolve at all, and (b) the divine promise that opens up the cosmos as well as human history and our personal lives to an ever-renewing future. In referring to Christianity's "revelatory image" henceforth I am including both of these intimately related themes. Of course, one might highlight other dimensions of revelation, but a theology of nature will find the themes of divine humility and divine promise quite

---

11. See Niebuhr, *Meaning of Revelation*, 69.

encompassing. It will take for granted also that both of these features are expressions of the infinite self-giving *love* that is the very essence of God (John 4:8).

Christian theology understands its revelatory image of mystery to have been given decisively in the person and career of Jesus, who for Christians is the fullness of the revelation of God. The one who has seen Jesus "has seen the Father" (John 14:9). In Jesus' obedience, crucifixion, and death, theology today, perhaps more than ever before, discerns the image of God's humble self-emptying—in other words, the divine *kenōsis*. The idea that God becomes small for the sake of relating most intimately to the creation has always been present, though often ignored, in Christian tradition.[12] What is revealed in the incarnation, passion, and crucifixion of Jesus is the paradoxically illuminating image of a vulnerable, suffering God who, out of love for the world, renounces any claims to "control" the course of events, and who gives the divine selfhood over to the entire universe as its silent but ever faithful font of renewal.

Christian teaching and preaching often tend to ignore the "descent of God" revealed in the crucifixion. Likewise, theologians and teachers fail to emphasize sufficiently that God is revealed paradoxically also as the mysterious power that opens the world to an always new future, as revealed in the biblical promises from Genesis to the Revelation of John, and especially in the Gospel accounts of Jesus' resurrection from the dead. The God revealed in Scripture is the one who makes all things new (Rev 21:5). But if God the Creator is also the world's Future, then it would not be surprising that the universe itself would have always had an anticipatory character. So today a theology of nature may connect the divine promise to what science now refers to as emergence, the tendency of the universe to give rise occasionally and spontaneously to new forms of complexity, especially in the spheres of life and mind.[13] When theology reflects on cosmic origins, the evolutionary thrust of

---

12. See especially Joseph Hallman's important but neglected book *The Descent of God: Divine Suffering in History and Theology* (Minneapolis: Fortress, 1991). Also Donald G. Dawe, *The Form of a Servant: A Historical Analysis of the Kenotic Motif* (Philadelphia: Westminster, 1963); Lucien J. Richard, O.M.I., *A Kenotic Christology: In the Humanity of Jesus the Christ, the Compassion of Our God* (Lanham, Md.: University Press of America, 1982); Jürgen Moltmann, *The Crucified God: The Cross of Christ as the Foundation and Criticism of Christian Theology*, trans. R. A. Wilson and John Bowden (New York: Harper & Row, 1974); and Hans Urs Von Balthasar, *Mysterium Paschale: The Mystery of Easter*, trans. Aidan Nichols, O.P. (Edinburgh: T & T Clark, 1990).

13. See Harold J. Morowitz, *The Emergence of Everything: How the World Became Complex* (New York: Oxford University Press, 2002).

life, and the emergent character of nature in general, the revelatory image that combines God's descent with God's promissory nature can frame natural phenomena in such a way as to disclose a deeper meaning in natural processes than science alone can ever discover. There is nothing here that conflicts with science, but there is a discernment that goes far deeper than science. The question of what the deeper meaning consists of as far as nature is concerned will be addressed in the following chapters.

At the risk of merely mentioning an issue that needs much fuller treatment, it is appropriate at this juncture to advert to the fact that issues of science and religion are inevitably going to differ when "other" religious traditions are brought into conversation with science. As this book proceeds, I hope it will become evident that the twofold revelatory image will not be completely foreign to the way in which some non-Christian religions also experience the grounding mystery of the sacred. Although the point of departure for Christian theology is the picture of Jesus as the Christ, other traditions, as they reflect on nature in the light of science and religious teachings, may also see the face of mystery presenting itself both as a *descent* implied in the intimate relationality of the divine mystery to the universe and as a *promise* that opens the world to a future that can in some way nourish the hopes of all people. In any case, I trust that religious readers of this book who may not be Christian will be able to find at least something of significance in the outline of the revelatory image that will dominate the present work. Let us now probe further into each aspect of this two-sided image.

THE HUMILITY OF GOD

In a Christmas sermon at Holy Trinity Church in Washington, D.C., Patrick F. Earl, S.J., recently gave the following reflections:

> We know the story of the birth in Bethlehem. We have seen portrayed in living pageants and in colorful statues the scene of the babe lying in a manger. Reflecting on this scene, the Jesuit poet, Gerard Manley Hopkins, spoke of "God's infinity dwindling—dwindling into an infant." "Dwindling" is the sign given us to recognize our savior. We will not find an infinite God—but an infant snugly wrapped and lying in straw. We have a dwindling—a descending and diminishing God—who comes to us in all the majesty of a feeding trough . . . Christmas—the feast of Incarnation— the feast of God becoming flesh of our flesh—this feast celebrates God's dwindling into us . . . We too are where God dwindles. We too are where

God takes on human flesh and life. It's in human flesh and life—in our fleshy selves—in our concrete lives where the divine light shines.[14]

That such a homily can be given in a parish church today is reflective of a general shift in Christian theological education toward emphasizing the theme of God's self-emptying, which lies at the roots of Christianity but which has not always stood out so explicitly. What could be added to Fr. Earl's sermon, however, is that the self-humbling occurs in God's relationship not only to humans but also to the entire universe. The very existence of the universe is the primary outcome of God's dwindling. The theme of the descent or humility of God is entailed by the christological and trinitarian teaching that God is one with the person and fate of Christ. Trinitarian theology (according to which each of the three persons participates fully in the life of the others) allows us to conclude that in Jesus' life and death what is revealed is nothing less than the *kenōsis*, that is, the self-emptying love of God.

However, the divine descent in no way means that God is weak or powerless. In Christ's passion God is presented to faith as vulnerable and defenseless, but, as Edward Schillebeeckx has remarked, vulnerability and defenselessness are more capable of powerfully disarming evil than all the brute force in the world could ever accomplish.[15] Think, for example, of the effectiveness of Mahatma Gandhi, Martin Luther King, and all the holy women and men who have accomplished so much through nonviolent witness and activity. "Power" means the capacity to bring about significant effects, but this does not necessarily require the external use of force. In the humility manifested in the manger and on the cross, God can be a power of attraction in a most unifying and creative way (John 12:32).

If some readers suspect that focusing on the theme of God's descent is only marginal to orthodox Christian faith, it is worth noting that the late Pope John Paul II, no theological radical himself, instructed theologians that their fundamental business is that of exploring the mystery of the divine *kenōsis:*

> The chief purpose of theology is to *provide an understanding of Revelation and the content of faith.* The very heart of theological enquiry will thus be the contemplation of the mystery of the Triune God. The approach to this mystery begins with reflection upon the mystery of the Incarnation of the Son of God: his coming as man, his going to his Passion and Death, a mystery issuing into his glorious Resurrection and Ascension to the right hand

---

14. See http://www.holytrinitydc.org/Homilies/hearl122405.htm.

15. Edward Schillebeeckx, *Church: The Human Story of God,* trans. John Bowden (New York: Crossroad, 1990), 90.

of the Father, whence he would send the Spirit of truth to bring his Church to birth and give her growth. From this vantage-point, the prime commitment of theology is seen to be the understanding of God's *kenosis*, a grand and mysterious truth for the human mind, which finds it inconceivable that suffering and death can express a love which gives itself and seeks nothing in return.[16]

The image of a self-emptying—and hence intimately *relational*—God, the absolute outpouring of goodness and love, is the very essence of the Christian experience of revelation. Reflection on the universe in light of this image, therefore, is not optional to a theology of nature. The self-giving of God is what the doctrines about creation, Christ, redemption, eschatology, and the Trinity are really all about at bottom.[17] Theology more than occasionally loses contact with the "irrational" revelatory notion that the ground of all being is endless, unbounded, self-giving love. But apart from the descent of God all the complex systematic theological tracts whose alleged purpose is to clarify revelation fall flat. They fail to illuminate what is going on not only in our own lives but also in the depths of nature. Hence theology, especially in its dialogue with science, must keep the mystery of a dwindling, humble, self-emptying God at the very center of its reflections. If it does so it may discover to its surprise that the recent scientific discoveries about the universe, especially about cosmic origins, evolution, and emergence, are not nearly so problematic theologically as they are when God is presented as an omnipotence without love, an intelligence without compassion, an absoluteness without relationality, an eternity purified of temporality, or an immutability sterilized of inner drama.

Even God's creation of the universe, which may at first conjure up images of fireworks on a grand scale, looks different if we think of it in terms of the divine humility rather than simply on the model of efficient causation. As theologian Jürgen Moltmann insightfully puts it:

> God "withdraws himself from himself to himself" in order to make creation possible. His creative activity outwards is preceded by this humble divine self-restriction. In this sense God's self-humiliation does not begin merely with creation, inasmuch as God commits himself to this world: it begins beforehand, and is the presupposition that makes creation possible. God's creative love is grounded in his humble, self-humiliating love. This

---

16. John Paul II, Encyclical letter *Fides et Ratio* (September 14, 1998), # 93. See http://www.vatican.va/holy_father/john_paul_ii/encyclicals/documents/hf_jp-ii_enc_15101998_fides-et-ratio_en.html.

17. See Eberhard Jüngel, *The Doctrine of the Trinity: God's Being Is in Becoming,* trans. Scottish Academic Press Ltd. (Grand Rapids: Eerdmans, 1976).

self-restricting love is the beginning of that self-emptying of God which Philippians 2 sees as the divine mystery of the Messiah. Even in order to create heaven and earth, God emptied himself of all his all-plenishing omnipotence, and as Creator took upon himself the form of a servant.[18]

An assimilation of the kenotic image of God may allow theological reflection to bring new meaning to nature in the midst of all the vexation aroused by scientific discoveries of the singularity of its Big Bang origin, the indeterminacy of quantum events, and especially the erratic character of biological evolution. Not only human freedom but also the emergent character of the entire natural world can be rendered more coherent than ever before when they are understood in terms of the theme of God's humility.

Until quite recently, theologians have been reluctant to take seriously this most shocking of revelatory images. Perhaps this is understandable. The picture of God's descent comes into the midst of a world whose thoughts about power and causation are not ready for the news that only a vulnerable love can be completely effective. So novel and surprising is this idea that enlightened minds are immediately compelled to doubt it. Yet the failure to locate a divine humility at the foundation of all being leads, I think, only deeper into perplexity about the way nature now appears to science.

Theology's failure to place the motif of God's descent at its very center leaves Christians at a disadvantage when it comes to finding meaning in the many previously unknown aspects of nature that science has been uncovering, especially over the last two centuries. I am referring here not only to the troubling way in which Darwin and his followers have shown that life struggles and diversifies on earth but also, at a deeper level, to the fact that the entire cosmos turns out to be a *story* rather than something essentially stationary. Our religious ancestors knew nothing about the fourteen-billion-year-old cosmic narrative that science has uncovered during the last century and a half. Most Christians traditionally assumed that the natural world was created by God in the beginning primarily to be a stage for the human drama. But today, as a consequence of recent scientific discovery, we realize that the stage itself is a grand drama and that our own human story is a very recent chapter in an immeasurably larger cosmic epic. I doubt whether Christians can make theological sense of these astounding new discoveries without connecting them to the central idea of God's descent.

Most Christians—along with religious people in general—still ignore the

---

18. Jürgen Moltmann, *God in Creation: A New Theology of Creation and the Spirit of God,* trans. Margaret Kohl (San Francisco: Harper & Row, 1985), 88.

twentieth-century discovery that the cosmos is a narrative still in process, that creation is not finished and that ambiguity, along with the possibility of tragic loss and suffering, is inseparable from any universe that is still in the making. One reason for this inadvertence is that the new scientific information does not accord well with certain inherited ideas of divine omnipotence. However, a theological focus on the revelatory image of God's humility (along with the theme of promise) can make abundant room for just the kind of universe that science is giving us. We do not need to invent a new theology in order to make religious sense of recent scientific discoveries. As we search for meaning in the universe of science, we need only recapture the "core hypothesis" of a self-emptying God who lovingly renounces any claims to domineering force and whose power cannot be severed from self-giving love. I shall develop this proposal more fully as the book moves along.

THE DIVINE PROMISE

In its attempt to understand what is really going on in the universe, a scientifically informed theology of nature will also take advantage of the promissory aspect of the biblical faith. As I have already pointed out, God's self-revelation becomes apparent to faith first in the form of a promise that opens up a new future.[19] The biblical account of God's gracious call of Abraham from out of the blue sets the whole tone for the main religions of the Western world. The God of Abraham is a God of promise. Because of this ancestry, Christianity, as Moltmann and others have recently recognized, is a religion of the future. Even Jesus' post-Easter appearances to his disciples are not so much theophanies as promissory events reminiscent of the call to Abraham to move into the great future laid open to him by God.[20] Promissory events are what brought Israel into being, and it is the intensely promissory events surrounding the appearances of Jesus to his disciples that gave rise to the Christian community and its reborn hope. The revelation of a momentous promise continues to give the gathering of believers, the community of hope known as the church, its fundamental reason for being. This hope must also be the foundation of a theology of nature. In other words, the whole universe may now be thought of as anticipatory, that is, of being already grasped

---

19. See especially Gerhard von Rad, *Old Testament Theology*, trans. D. M. G. Stalker, 2 vols. (New York: Harper & Row, 1962-65); and Jürgen Moltmann, *Theology of Hope: On the Ground and the Implications of a Christian Eschatology*, trans. James W. Leitch (New York: Harper & Row, 1967).

20. Moltmann, *Theology of Hope*, 139-229.

by the futurity of the divine mystery that comes to awareness in biblical traditions.

In its being shaped by promise Christianity links up, at least in a broad sense, with other religious traditions that aspire to final liberation from all that imprisons the world and life. So our reflections on the Christian understanding of revelation need not be isolated altogether from the wide plurality of religious experiences of a mystery that promises redemption. Because of its strong incarnational and sacramental emphasis, however, Christianity is especially obliged to keep the cosmos and its future squarely in the forefront of theology, even though it has not always done so. In taking on human flesh, God gathers the entire universe into the divine life, and because of Christ's and our own inseparability from this universe there can be no personal liberation *from* the cosmos, but only *with* it. Salvation, therefore, must mean much more than a harvesting of souls from the material world. The physical universe also needs to be saved if Christianity is true.

Because of developments in science, especially cosmology and evolutionary biology, we now realize that human history and the history of salvation can occur only within the more encompassing setting of the cosmos.[21] So it has to be toward the whole universe and its fulfillment that Christian hope extends. This cosmic perspective must also be wide enough to prepare theology for any possible future encounters with intelligent beings in other parts of the universe. Yet, even aside from that prospect, a cosmic perspective can make room here and now for our respecting nonhuman forms of life on our own planet so that they too may have a future. In doing so, a theology of nature informed by science can enlarge our sense of mystery and the impact of revelation.

Since we live in an age of astrophysics and evolutionary biology we have been gifted with a new picture of the universe. It is no longer fruitful to insulate theology from what the sciences are revealing. Can theology persist in dissociating the idea of revelation from the grandeur of the cosmic narrative that scientists are now putting on exhibit? Unfortunately, with some scattered exceptions, theology has so far virtually ignored this dazzling display. Thus it has failed to comment on some of the questions that are most intriguing to informed people today. It has scarcely noticed that the perennial human concern about origins, meaning, destiny, and human obligation is being expressed with no less urgency than ever, though in a wider setting: If we are part of an evolving world where is this world heading? Does the universe have a purpose? Does cosmic evolution have any direction to it? What are our own obli-

---

21. This is a major theme in Jürgen Moltmann's book *God in Creation*.

gations in such a world? How does our species fit into the evolutionary picture? Is cosmic emergence a futile thrust into the void or perhaps a response to an invitation to journey into the mystery God? If the universe is heading toward eventual decline and death, what is the point of our own lives? What sense can we make of the randomness, indifference, and impersonality of evolution by natural selection? What is the meaning, if any, of deep cosmic time? If the world is created by God, why did it "fool around" for so many billions of years before bringing forth conscious beings? What happens to religion, and particularly to belief in the redemptive significance of Christ, if intelligent and spiritual beings exist elsewhere?

THE TASK OF A THEOLOGY OF NATURE

A theology of nature should address these and similar questions that scientifically informed people are asking. Not all of them can be given full coverage in the present book, of course, but my point is that they should not remain off-limits to theological reflection on the meaning of revelation. The universe attracts a great deal of attention today, but theological reflection seems at times to take pains to avoid it. I suspect that behind this indifference theology still harbors a residual dualism, an otherworldliness or a kind of cosmic pessimism of its own. Throughout the modern period theologians have been content for the most part to hand over to science the task of understanding nature, while they have retreated into preoccupation with issues of personal and social concern. The latter are worthy issues too, of course, but if theology fails to respond to the largest of human questions—those having to do with the meaning of the universe—it will seem increasingly irrelevant to those who appreciate the vistas of scientific discovery.

As Galileo insisted several centuries ago, religion and theology have no business dishing out information that human intelligence can find on its own. Nevertheless, theology's business *is* that of addressing the larger questions that arise when enlightened inquirers find themselves wondering about the meaning of what science has observed. Always respecting the autonomy of science, a theology of nature may still ask whether the emergent universe of contemporary science can be fruitfully situated within the wide circle of meaning evoked by the revelatory images of God's descent and promise. A theology of nature must never give the impression of intruding into the work of scientific investigation. But it lies within theology's proper sphere of concern to connect the substance of scientific understanding of nature to faith's sense of a self-giving mystery that opens up the world to an ever-new future. Situating scientific results within a revelatory worldview can even have the effect of liberating science from the materialist quagmire into which naturalists continue to dump the data of scientific research.

A theology of nature is not indifferent to the individual's personal search for meaning. Revelation must speak to each of us in our solitude and our social existence as well. Here, however, my efforts will be directed toward connecting the revelatory image of Christian faith with the universe of the natural sciences. This perspective will allow us to deprivatize revelation and reach even beyond its sociopolitical meaning. Of course, an awareness of God's self-revelation first occurs in Jesus' intimate *personal* experience of God as *Abba*, but a full examination of the meaning of revelation carries our concern for its application eventually outward into the wider universe. Jesus and his work of redemption cannot be isolated from the web of natural relationships that tie him and us into the cosmos and its history. Our being is essentially cosmic as well as communal, and even in our aloneness each of us is tied to the universal. A theology of nature will keep in mind the four infinites—the immense, the infinitesimal, the complex, and the future—of nature. And it will speak especially to those questions that have to do with this mystery-encompassed universe from which we are inseparable.

Among such questions today those raised by our global environmental situation are paramount. For many people it is proximately the current global ecological predicament that leads to the most theologically momentous limit questions about the universe: What is the universe all about? Why should we bother to care for our little corner of it? If nature seems finally indifferent to life, why should we be worried about conservation? What are our obligations to the universe that bears us along?

Can the revelatory image of God's descent and futurity shed any light on these and similar issues? A theology that focuses exclusively on personal and social issues is ill-equipped to do so. The acosmic leaning of traditional theology deserves the accusation that "revealed" religions are responsible for promoting a noxious idea of cosmic homelessness that sets humans spiritually adrift in the universe, leading us to be indifferent toward nature. A theology of nature will take into account and respond to such criticism. Here again, theology cannot dictate specific environmental policies any more than it can lay out definitive social or economic programs. Nevertheless, it may attempt to answer the limit question that asks why we should bother to take care of our natural environment at all. The revelatory image of a self-humbling and promising God may help shape a vision of reality that will help us address this urgent concern.[22]

---

22. I do not have the space to address the question of revelation and ecology in the present book.

SUMMARY AND CONCLUSION

The present book asks about the meaning of the Christian revelation for our understanding of nature. It presupposes the framework of the new cosmology that arises out of geology, biology, physics, astrophysics, and cosmology. It assumes the fundamental correctness of evolutionary biology and other firmly established discoveries of modern science. If scientists and other intellectuals are going to pay attention to a theology of nature, it is imperative that the understanding of revelation presented to them is in no way contradictory to the most up-to-date versions of contemporary science.

Over the past century and a half Christian theology has lived and moved most comfortably in the context of questions about the meaning of history or personal existence. Only tangentially has it concerned itself with cosmology. In its earliest expressions religious symbolism clung closely to the rich vine of nature, usually without being self-conscious about it. Even today most religions retain at least vestiges of prescientific cosmologies. Meanwhile, however, modern science has altered our sense of the cosmos and produced new pictures of nature. In the face of such a transition, Christian theology has not always kept pace, and it has sometimes attempted to de-cosmologize faith completely. Often it has left the universe—and that includes our own evolutionary heritage—out of its visions of reality.

Over a half-century ago, for example, the Protestant biblical theologian Rudolf Bultmann argued that revelation is essentially God's address to the hidden inner freedom of each person. As far as Bultmann could see, the non-human natural world has little to do with the Gospels. God's relevance consists of giving us personal freedom in Christ, and this gift ideally carries over to improving our social life as well. But Bultmann made little effort to relate his existentialist understanding of redemption to the scientific view of nature with which he was quite familiar. He rightly saw the need to demythologize the scientifically obsolete expressions of Scripture, but he saw no need to *re-cosmologize* theology in terms of contemporary science. As a result of this kind of evasiveness—which Bultmann was not alone in practicing—theology has severed nature from the promissory side of revelation.[23]

In the biblical traditions a divine promise comes through Abraham to a whole nation and eventually to all people. Since nature and history are inseparable in the biblical understanding, we are not stretching the idea of God's

---

I have attempted to do so, however, in *The Promise of Nature* (Mahwah, N.J., and New York: Paulist, 1993).

23. Rudolf Bultmann, "New Testament and Mythology," in *Kerygma and Myth*, ed. Hans Werner Bartsch, trans. Reginald Fuller (New York: Harper Torchbooks, 1961), 1-44.

promise too far if we now allow it to embrace the whole universe and the long history of nature unfolding through time. As we look at the various levels of depth to which science has brought our understanding of the cosmos, perhaps we may find at the bottom of it all the same divine promise that opened up a new future for Abraham, the prophets, and Jesus.

SUGGESTIONS FOR FURTHER READING AND STUDY

Delio, Ilia. *The Humility of God*. Cincinnati: St. Anthony Messenger Press, 2005.
Hallman, Joseph. *The Descent of God: Divine Suffering in History and Theology*. Minneapolis: Fortress, 1991.
Niebuhr, H. Richard. *The Meaning of Revelation*. New York: Macmillan, 1960.
Rahner, Karl. *Foundations of Christian Faith*. Translated by William V. Dych. New York: Crossroad, 1978.

4

What's Going on in the Universe?

Thou hast made thy promise wide as the heavens.

—Psalm 138:2

IN CHRIST THE ULTIMATE MYSTERY that encompasses all created being is revealed not only as self-giving love but also as saving future. What then should we expect the universe to look like in light of the promise that enfolds it? Science has demonstrated over the past century and a half that the universe is an ongoing process that is unfathomably vaster and older than we had ever imagined before. The cosmos came into being billions of years before the arrival of human history, Israel, and the church. Apparently God's creative vision for the world stretches far beyond what transpires in terrestrial and ecclesiastical precincts. Nevertheless, a Christian theology of nature has to wager that the promissory perspective of biblical faith that first came to light in a tiny nation on a very small planet not very long ago in cosmic time is applicable to cosmic reality in all its enormous breadth and depth. The promise of God is "wide as the heavens," and if the heavens seem immensely wider today than ever before, so also should our sense of the reach of God's promise.

Thus, the long cosmic epochs that preceded the emergence of humanity, Israel, and Christianity must also be interpreted by faith as having always been touched by the futurity of God. In the Bible there is no separation of nature from the history of promise. With eyes of hope inherited from the faith of Abraham, Christians are encouraged to look for signs of a salvific future opening up amid all the ambiguities of natural, not just human, history. Today this may prove to be less difficult than ever before, since the latest developments in natural science now show that the universe has always had an emergent or *anticipatory* character. Even in the midst of the perpetual perishing that is the lot of all created being, the universe, it seems, has never been closed off to surprising future outcomes. From its beginning fourteen billion years ago it has continually made room for new and unprecedented achievements. It still does, especially through one of its most enthralling evolutionary inventions, the restless hearts and minds of human beings.

Once the phenomenon of mind had burst onto the terrestrial scene, the world's posture of straining toward the future began to take the form of religious aspiration everywhere. In the world of Israel this longing broke through in a new surge of hope for the future, a disposition that shaped the religious awareness of Abraham, the prophets, Jesus, and the earliest Christian communities. This habit of hope is not just human imagining. From a cosmological point of view it is the way in which the universe, of which we are fully a part, opens itself to new creation up ahead. Through our own expectant posture the cosmos still continues to scan the horizon for the dawning of a mysterious and elusive new future. It is from this future that we still look for the coming of God and new creation.

And yet, these days many people still wonder how Christian hope for cosmic redemption could ever be reconciled with what science has to say about nature. A scientific perspective understands the universe to be hidebound by deterministic laws. No room seems to be available for a truly new future ever to occur. Natural science, because of its method of understanding present phenomena in terms of earlier and simpler lines of causation, can know nothing of any promise of fulfillment. Nor is it supposed to. When science makes predictions, it does so only on the basis of what it already knows. Its pictures of the future are extrapolations from the invariant routines known as physical laws.

Take, for example, the Second Law of Thermodynamics. It holds that the most probable future state of the universe is one in which the available energy to produce and sustain such remarkable emergent phenomena as life will become irreversibly lost. The universe is subject to entropy. Like a clock spring winding down over the course of time, it keeps losing the power to do work, including most notably the building up of organic complexity. And science knows of nothing that will wind it up again. At some point far away in the future the universe that still holds an enormous reserve of energy will expire completely. On the basis of what scientists are aware of at present, it will eventually arrive at a state of energetic paralysis. This ending will cause life, consciousness, and every other significant outcome of natural and historical processes to disappear forever from the smoldering remains of the cosmos.

Does this prospect perhaps render hope futile and the universe purposeless? Numerous scientifically educated writers during the past century have concluded that it does indeed. The British physicist James Jeans, for example, claimed that science has uncovered a universe hostile, or at least indifferent, to life and humanity. The cosmos is destined for final exhaustion from the thermodynamic disease of entropy. Jeans asks, therefore, whether there is any more to human life than strutting for a brief moment "on our tiny stage," armed only "with the knowledge that our aspirations are all but doomed to final frustration, and that our achievements must perish with our race, leaving the universe as though we had never been?"[1] Echoing the same pessimism, the renowned philosopher Bertrand Russell (1872-1970) provides an oft-quoted lyrical testimonial to the final futility of human life in a cosmos destined for utter ruin:

> Brief and powerless is Man's life; on him and all his race the slow, sure doom falls pitiless and dark. Blind to good and evil, reckless of destruction, omnipotent matter rolls on its relentless way; for Man, condemned today to lose his dearest, tomorrow himself to pass through the gate of darkness, it remains only to cherish, ere yet the blow falls, the lofty thoughts that ennoble his little day; disdaining the coward terrors of the slave of Fate, to worship at the shrine that his own hands have built; undismayed by the empire of chance, to preserve a mind free from the wanton tyranny that rules his outward life; proudly defiant of the irresistible forces that tolerate, for a moment, his knowledge and his condemnation, to sustain alone, a weary but unyielding Atlas, the world that his own ideals have fashioned despite the trampling march of unconscious power.[2]

THEOLOGY AND COSMIC PESSIMISM

What response can theology make, then, to this placid certainty that the cosmos will indeed eventually perish—and, along with it, all traces of life and culture. At present, one must admit, there is no scientific reason to expect that the universe can avoid an eventual death by entropy. But to theology this should be no surprise. Is it not true that every finite reality, including the expansive and long-enduring set of things we call the universe, is still limited in space and time? Theology has generally thought this to be the case. To be finite, after all, is to be subject to the threat of nonbeing, and the physical universe cannot be an exception.[3] Only a new creation can save the universe, and it is for this, not the indefinite prolongation of the present cosmos, that Christianity hopes.

Nevertheless, it is toward a new creation of *this* universe, not its replacement by another, that Christians look. If there is a reason for our hope, there cannot be a complete discontinuity between what is going on in the cosmos

---

1. James Jeans, *The Mysterious Universe,* rev. ed. (New York: Macmillan, 1948), 15-16. The book was first published in 1930.

2. Bertrand Russell, *Mysticism and Logic and Other Essays* (New York: Longmans, Green, 1918), 46ff.

3. Paul Tillich, *The Courage to Be* (New Haven: Yale University Press, 1952), 32-39; and *Systematic Theology,* 3 vols. (Chicago: University of Chicago Press, 1963), 1:209.

right now and any final, redeemed state of things. In a very deep sense the present perishable world must matter eternally to Providence, and so it is proper to look for signals here and now, perhaps very subtle ones, that this may be so. Theology, it seems, is obliged to show how a perishable cosmos need not be a meaningless one.

However, to do so theology would first have to assume that there is a sense in which perishing is not absolute. Theologically speaking, nothing finite could be purposeful unless it partakes of the eternal, even in the midst of its perishing. Christians are encouraged to hope, therefore, that everything that has ever happened or will happen in the universe is taken into the compassionate care of God. In an ever intensifying relationship to God all things that perish, including the whole course of events we call the universe, can be transformed into a beauty beyond imagining. "We hope to enjoy forever the vision of Your glory . . ."

But isn't such hope too much of a stretch, especially in an age of science? Once again, science itself, because of its orientation toward earlier and simpler lines of causation that can establish only what is "probable" in the future, cannot promise any such fulfillment, nor should we expect it to. If science were our only way to true and complete understanding, we would have to conclude that sheer nothingness awaits the cosmos. However, science by definition sketches only a limited picture of what is really going on in the universe. It follows Occam's razor, the axiom that one should not resort to multiple or complex explanations when a single or simpler one is available. Scientists are instructed to explain presently visible phenomena on the basis of fixed laws, in terms of what has already happened in the causal past, and in the currency of elemental units such as atoms, molecules, cells, or genes.

Scientifically speaking, this way of looking at nature is unobjectionable— as long as scientists acknowledge the inherent limitations of their method. A reductive method, one that employs simplifying mathematical models to represent complex entities, is essential to scientific understanding. Reductive generality has proven to be insightful as well as technologically fruitful. Scientific method, however, is not equipped to reach into the ultimate depths of nature or the world's open future. It cannot predict with full precision the genuine novelty that will arise today, let alone what will happen in the cosmic future. When science sees something new and remarkable in nature, its habit of mind is to show that, at bottom, it is really an instance of what is old and unremarkable. Every apparently new occurrence is at bottom an expression of unbending physical laws that apply everywhere and always. Life, for example, is an instance of chemical processes already functioning in nonliving stuff. To a reductive method there is nothing really new about living organisms since the entire realm of living beings is simply an interesting application of invariant physical and chemical laws that were operative in cosmic history long before the first living cells emerged. If it can be shown that biology is ultimately reducible to chemistry and physics, as Francis Crick and many others have claimed, then what appears newly emergent in life will be exposed as a mask behind which there exists only a humdrum physical simplicity. All the living extravagance and versatility brought into being by the twists and turns of evolution, according to the Oxford physical chemist Peter Atkins, are mere simplicity masquerading as complexity.[4]

Of course, since they are human persons like the rest of us, scientists are captivated tacitly by the puzzling fact of emergent new phenomena in cosmic history. But no sooner do they observe that life, mind, and other emergents are "interesting" or "remarkable," than they try to suppress their surprise by explaining the later-and-more in terms of the earlier-and-simpler. The dimension of futurity is ignored. Thus, everything that occurs in nature's evolution is taken to be a kind of façade behind which one will eventually discover the unchanging laws of physics and chemistry that have been running on in the same way unceasingly. In all the wide array of novel cosmic creations, the underlying constants of nature go on functioning as always. Novelty, therefore, must be an illusion.

However, this denial of novelty, and along with it the subversion of hope, is not a conclusion that science itself can legitimately reach. Rather, it is a core doctrine of the beliefs known as scientism and scientific naturalism. Truly thoughtful scientists willingly acknowledge the self-limiting methods of their various disciplines. They realize that every distinct scientific field can make progress only by leaving out, or abstracting from, what other disciplines cover. But to the scientific naturalist there is nothing real that cannot be reached and fully explained by science. This belief entails a suppression of what I have called the fourth infinite—that of the always open future—which in a Christian worldview provides the space for faith, hope, and cosmic emergence yet to come.

Theological reflection rightly opposes naturalistic pessimism, therefore, and it concurs with the more thoughtful scientists who are fully aware that their method of inquiry inevitably leaves out a lot. Science does not have any access, for example, to the realm of *subjectivity*. Scientific method, which idealizes publicly accessible understanding, cannot get inside the sentient phenomena in nature. It has no direct access to those centers of experience that allow certain beings to feel or become conscious of their environments. Sci-

---

4. Peter W. Atkins, *The 2nd Law: Energy, Chaos, and Form* (New York: Scientific American Books, 1994), 200.

ence cannot even say what it means for each atom, molecule, cell, or organism to be itself. All real individuality dissolves in the acids of scientific generality. Scientific hypotheses, theories, and laws inevitably abstract from the uniqueness of every entity or event. Faith, on the other hand, sets before us a God concerned with the "thisness" of everything.

To give the most obvious example of science's limited scope, we need only notice how little it has to say about the immediate awareness we each have of our own inner life, feelings, thoughts, and aspirations, including our desire to understand and know. This is the world of human subjectivity. Science cannot reach into this world and lay it out in an "objective" way. If it tries to approach the phenomenon of "thought" or "mind" in an exclusively objective way, science will inevitably pass over the "insideness" of our existence. This is nothing for scientists to be embarrassed about. The fact that science can "see" only objects and not subjects is simply part of the definition of science. To deny that subjectivity is real, however, is a materialist dogma, not a scientific conclusion.

Moreover, humans are not the only subjects existing in the web of nature. Animals also have incommunicable private feelings, desires, and the capacity to enjoy life or to experience fear and pain. They have sentient awareness and even a kind of consciousness. So it would be egregiously anthropocentric to deny, as did the philosopher René Descartes, that there is a vein of unreachable interiority in most living beings. The philosopher Alfred North Whitehead even maintains that *every* actual entity in nature, including the energy events that make up the purely physical world, is a kind of subject. All things, he insists, synthesize internally their environments in a way that science cannot grasp. Even though rocks and other aggregates are devoid of feelings, they are all made up fundamentally of subjective "occasions of experience." There is a kind of responsiveness in every actual entity, and this is what allows God to have a persuasive influence on nature.[5] Beyond science's restrictive purview, natural process has always had a capacity to feel and respond to the call and coming of God, and this is ultimately why the universe has never been able to stand still.

If such a philosophical position seems too poetic, it is sufficient for present purposes to take note only of the fact that at least some beings in nature have a side to them that science cannot catch and objectify. Since we humans are fully part of nature, as Pierre Teilhard de Chardin has emphasized, the fact of our own capacity to have subjective experiences is already enough of a rent or

---

5. Alfred North Whitehead, *Process and Reality*, corrected ed., ed. David Ray Griffin and Donald W. Sherburne (New York: Free Press, 1978), 23, 25, 157, 221.

tear in the fabric of nature to warrant our attributing to the universe a kind of "withinness" that renders it resistant to complete scientific objectification.[6] Science simply cannot extract and render publicly visible the subjectivity that has gotten into nature. So frustrating is this slipperiness to extreme naturalists that some of them even deny that subjectivity has any reality at all, even in us humans.[7]

In order to retrieve what passes through the wide-meshed net of science, I propose that there is need for a "wider empiricism" to supplement scientific experience and understanding. By a wider empiricism I mean a way of seeing or experiencing that is sensitive to the insideness of things as well as to the genuine novelty that emerges, sometimes explosively, in natural history. What conventional science sets before us is by no means everything that is going on in nature. Therefore, a theology of nature that tries to be in touch with what is *really* going on in the universe must take into account aspects of nature that are not available to scientific method. The wider way of seeing that I will rely on is not opposed to science, but it puts the inquirer in touch with dimensions of the universe that scientific method cannot. My proposal is that if we deploy *all* of our perceptive apparatus, which science does not, we may be able to make out, at least vaguely, a purpose and promise at the deepest layers of our perishable universe. Our hope is empirically justifiable after all. More on this later.

LAWFULNESS AND INDETERMINACY

For the moment, I want to clear up a possible misunderstanding of what I am saying. Even though science, at least as it has been understood since the seventeenth-century, is not equipped to contact the realm of subjectivity and is likewise unable to predict the unique shape of real novelty that will emerge in the future, this is no reason to assume that future cosmic happenings will somehow contradict the laws of physics, chemistry, and biology. The predictable habits of nature will go on functioning as before. It is just that they will be taken up into an indeterminate array of novel configurations. Even what Christian tradition refers to as "miracles" need not be thought of as altering the regularities of nature.

---

6. Pierre Teilhard de Chardin, *The Human Phenomenon*, trans. Sarah Appleton-Weber (1959; Portland, Ore.: Sussex Academic Press, 1999), 23-24.

7. E.g., Paul M. Churchland, *The Engine of Reason, the Seat of the Soul: A Philosophical Journey into the Brain* (Cambridge, Mass.: MIT Press, 1995).

A comparison of science with grammar may help to clarify this point. The unbendable rules of grammar are operative throughout the writing of this book. These rules place constraints on every sentence and paragraph you will be reading. Yet the actual content of what I will be saying in the pages ahead is not determined by the rules of grammar. Grammatical rules *constrain* but cannot account fully for what I shall be writing. There will be plenty of room for the author to say new and strange things further on without violating, for example, the ordinance that a predicate must agree in number with the subject of a sentence. Indeed, the expression of new ideas cannot occur without my adhering consistently to rigorous grammatical norms.

You may assume, then, that the rules of grammar will still be operating to organize what I will say later on in this book just as they are operating on this page. But grammatical expertise alone cannot give you any insight into the actual content of what I will be saying. Likewise there is plenty of room in the universe for new things to occur without ever violating in the slightest the laws of physics, chemistry, or natural selection. Scientists can justifiably predict that these laws will be operating in all future phases of nature's unfolding, but they cannot specify precisely what the actual emergent chapters of cosmic process will be like as they unfold. There is abundant latitude in our processive universe for new things to happen, no matter how rigorous the laws of physics may be. And if there is room for novelty there is also room for hope.

Science and Christian hope for new creation, therefore, are not in principle opposed to each other. When science understands events in terms of general physical laws, this should not be taken to mean that everything that will happen in the future is already preordained. Science's exposition of invariant physical laws is comparable to the "discovery," let us say, of the rules of syntax that underlie human language. For centuries people spoke and wrote intelligibly without having any formal knowledge of the exciting discoveries about syntax made by linguistics experts such as Noam Chomsky. Language automatically conforms to consistent rules, but knowing the rules will not give Chomsky any command over the actual content of what you or I will be saying.

So also this book will go on for chapter after chapter in accordance with familiar linguistic norms and grammatical rules, but its content will be different from that of any other book. So it would be inappropriate to burden the grammarian with the task of telling me what I should be saying. Likewise it would be wrong to load science down with the job of deciding whether hope for a meaningful outcome to cosmic process is realistic, or whether the universe is the carrier of a meaning. Nature will run along age after age adhering without exception to physical laws, but it will still be open to unpredictable outcomes.

A theology of nature, motivated by trust that the universe is grounded in the promises of God, is not opposed to science, and it must remain abreast of science. But science, in spite of what naturalists maintain, cannot by itself tell us everything about what is *really* going on in the universe. Even though "what is really going on" is not happening in violation of scientific laws, we should not expect science to take us down to the deepest strata of cosmic reality. What science is telling us about the natural world does place constraints on what theology can say about the universe, and scientific discoveries must be embraced by theology.[8] But theology is not contradicting science when it professes that nature, not just history, is a purposeful process.

CAN THE UNIVERSE HAVE A PURPOSE?

Before modern times, most people assumed that the universe had a purpose. The natural world existed for a reason, although it was not always easy to say exactly what this reason was. Philosophies and religions were aware of nature's flaws, of course, but they still viewed the cosmos as a "great teaching."[9] Sometimes they even saw it as a sacred text that could be read in depth by minds and hearts that were properly prepared. The book of nature held a deep message hidden from ordinary awareness. Great visionaries could sense that something important was going on in the cosmic depths, and their visions allowed others to experience the universe as purposeful. Pythagoreans, for example, heard a musical kind of harmony sounding forth from the bosom of nature. Ancient Israelites read the universe as an expression of divine wisdom. Egyptians thought of nature as encompassed by a realm of norms known as *Ma'at*. Indians subjected their minds and hearts to *Dharma*, and Chinese philosophers to the *Tao* that was the hidden and humble foundation of all being. Stoics read the cosmos as the outward manifestation of an inner rationality they called *Logos*. And the Gospel of John pierced beneath the surface of immediate experience to an eternal *Word* that was in the beginning with God, and that was God. To Christians the doctrine of the Trinity still expresses the intuition that beneath the surface of all things an unfathomable drama involving the distinct functions of three divine persons goes on forever and that whatever transpires in the finite universe is taken permanently into this mystery of love and creativity.

---

8. Holmes Rolston III, *Science and Religion: A Critical Survey* (New York: Random House, 1987), 26.

9. Jacob Needleman, *A Sense of the Cosmos* (New York: E. P. Dutton, Inc., 1976), 10-36.

Traditionally, almost all religions and philosophies have read the universe as the outward expression of a deep inner meaning. But in order to receive any message emanating from the cosmic depths, one had to have been prepared by undergoing a process of personal transformation. Being clever was not enough. Without undergoing a training or discipline one could not expect to understand what is really going on in the universe. In our own times a good science education allows one to read the cosmos, but only in a limited way, since science does not by itself change our hearts or values. Science has shown us nature's atomic and molecular alphabet, its genetic lexicon, and its evolutionary grammar. But it has not taught us how to read the universe, that is, how to access the deepest content of what is being written there. Scientific information keeps piling up, but the wisdom to make sense of it all does not keep pace.

Scientific naturalists, in fact, do not see the cosmos as a text bearing any deep meaning at all. To many of them, the universe is at heart a mass of meaningless matter on which the sheen of life and human history glimmers only for a brief cosmic moment. At best nature is a canvas on which humans may inscribe their purely human meanings, and in the end these fictions will be swallowed up by the same void out of which the universe is held to have unintelligibly appeared. As the above quotation from Bertrand Russell illustrates, the idea that something of lasting significance is happening in the depths of nature seems preposterous. Any religious proposal that the universe is here for a reason is not a topic for scholarly discussion, especially in view of the randomness, struggle, and general impersonality of nature as pictured by Darwinian science. The only message in evolution, says the philosopher Daniel Dennett, is that "the universe has no message."[10]

Is the perishable universe bringing about anything of lasting value? Theological responses to this question cannot ignore the prospect of an eventual physical death of the entire universe as well as the troubling features in evolution. I shall say much more about these disturbing realities respectively in chapters 6 and 9. For now I want only to reflect on two items in the current store of scientific information that make it at least conceivable that something of great significance is going on beneath the surface of nature. These are features that can easily be mapped onto the revelatory theme of divine futurity and promise. The first is that the physical universe is still in the process of coming into being. The second is that what has come into being already includes such an intensity of beauty that nature may be read as a great prom-

---

10. Daniel Dennett, interview in John Brockman, *The Third Culture* (New York: Touchstone, 1995), 187.

ise of more being and value up ahead. If the first fact reminds us to be realistic and not expect too much here and now from an unfinished universe, a taste of the second allows us to hope in a power of renewal that can eventually bring new birth to the whole universe. Let us look at each of these two aspects of nature more closely.

1. *The universe is still coming into being.* Evolutionary biology, geology, and cosmology have now established as fact that the cosmos is still emerging and that it remains incomplete. It is a work in progress, a book still being written. The incontestable fact of an emergent, unfinished universe may not seem to be much of a footing on which to erect a sense of cosmic meaning or find a good reason for hope, but at least it invites us to keep on reading. For if a wondrous plot is still unfolding beneath our feet and over our heads, we cannot expect its inner meaning to be fully manifest yet. Any purpose the universe may have will be at least partially hidden from our view—at least for now.

Hope is still possible in such a universe. As with any book in progress, we cannot yet read the universe all the way down to its ultimate depths, whether we are looking at it through science or theology. Its vistas are too large to permit such a grasp. However, we may still feel ourselves being swept up in the story, and in such a way as to experience the emerging cosmos with a spirit of expectation. The cumulative impression one gets from recent scientific discoveries is that the unfinished universe is continually greeted by an inexhaustible reservoir of freshness, which as the poet Gerard Manley Hopkins puts it, lives "deep down things."[11] The universe, both in its monumental temporal and spatial scale, but also in its extravagant diversity, fairly bursts with creative novelty. Nature can seem indifferent at times to individual organisms striving to adapt and survive. But even as it relies on a monotonous substrate of physical and chemical invariance, it is still open to being taken up into a fascinating and unpredictable tale of suspense. It is the universe's narrative character that will allow us, as we move through the chapters ahead, to connect science more and more intimately to the revelatory themes of divine descent and promise.

2. *The universe is the story of an unimaginably wide display of beauty.* By beauty I mean the harmony of contrasts, the ordering of complexity, the fragile combining of what is new with what is stable, of fresh nuance with persistent pattern.[12] The still-developing cosmos, we now know, has made its way gradually

---

11. Gerard Manley Hopkins, "God's Grandeur."
12. Alfred North Whitehead, *Adventures of Ideas* (New York: Free Press, 1967), 252-96; idem, *Process and Reality*, 62, 183-85; idem, *Modes of Thought* (New York: Free Press, 1968), 57-63.

from primordial radiation, through the emergence of atoms, galaxies, stars, planets, and life, to the bursting forth of sentience, mentality, self-consciousness, language, ethics, art, religion, and now science. Such emergence has not been without setbacks, but by any objective standard of measurement, it is gratuitous to deny that something momentous has been going on in the universe. The cosmos has at the very least busied itself with becoming *more* than what it was. What has emerged so far is not just a masquerade concealing an underlying physical simplicity. For all we know, the universe of today may itself be an early chapter in a deeply meaningful story that is still far from having been fully told.

So how are we to read a cosmic narrative that is still being written? Toward what ending can we see the story heading? These are appropriate questions for a theology of nature. Although they admit of no easy answer, they will stay with us throughout the pages ahead. But even at this point we may find it remarkable that the universe eventually abandoned the relative simplicity of its earliest moments and flowered, over the course of billions of years, into an astounding array of emergent complexity and diversity, including human consciousness with its moral and religious aspirations. There is plenty of room for wonder here. Something other than just the mere reshuffling of atoms has been going on in cosmic history. And while the journey from primordial cosmic monotony to the intense beauty of life, mind, and culture is no hard proof of an intentional cosmic director, this itinerary is at least open to the kind of "ultimate" explanation that a theology of nature rightly seeks to articulate. The universe, in any case, has had an overarching inclination to make its way from trivial toward more intense versions of beauty.[13] Purpose, as I understand it, means the working out or actualizing of something of self-evident value, and beauty certainly qualifies as such. Cosmic purpose consists, at the very least, of an overall aim—not always successful, but nonetheless persistent—toward the heightening of beauty.

But what about the dark side of things—the loss of life, the struggle and pain in evolution, and the moral evil pertaining to human existence? And what sense can we make of the dismal scenarios that cosmologists are now entertaining about the eventual, although certainly far off, demise of the universe? How do we know that all things will not finally trail off into lifeless and

---

See also Charles Hartshorne, *Man's Vision of God* (Chicago and New York: Willett, Clark, 1941), 212-29.

13. This aesthetic directionality was enough finally to convince the great philosopher Alfred North Whitehead, after a long period of agnosticism, that there must indeed be a profound point to the universe; see *Adventures of Ideas*, 252-96.

mindless oblivion? Organisms all die, and great civilizations sooner of later decay. Hence, if there is any purpose to the universe, perishing must be redeemable—not only our own, but *all* perishing. There has to be a permanence in the depths of the world process that redresses the fact that nothing lasts. Beneath the transient flux of immediate things there must be something that endures everlastingly, and in whose embrace all actualities attain a kind of immortality. Only a kind of cosmic redemption could finally justify our hope.

It is this hope that religions seek to express—in many different ways. Although religions are imprecise and inconsistent, their visions may be able to penetrate deeper into the essence of the universe than can the lucid abstractions of science. Science may be able to deal with the surface of nature, but the Christian intuition has always been that we shall find beneath the temporal flux of finite being an everlasting redemption, a "tender care that nothing be lost."[14] In God's experience, the entire sweep of events that we call the universe is endowed with permanence along with purpose.[15] Even the final dissolution of our own expanding universe does not necessitate the meaninglessness to which cosmic pessimists consign it. If its history, down to the last detail, can be forever internalized in the life of God, the universe will not be pointless after all.[16]

Today a theology of nature needs to extend faith's trust in God's care far beyond the terrestrial and human spheres to the totality of cosmic being. The God of revelation, therefore, is not only the promise-maker who summons the world to exist and arrive at ever more intense beauty; not only the humble and compassionate co-struggling and co-suffering companion as revealed in Christ; not only the tender care that preserves everlastingly all the transient value that emerges during the becoming of nature; but also the one who, through the Spirit of life, continually renews the face of creation.

In the context of contemporary science, as I shall continue to emphasize, any distinctively Christian theology must think of God as having the breadth and depth of feeling to take into the divine life the entire cosmic story, including its episodes of tragedy and its final expiration. Within the embrace of a self-humbling God, the whole universe and its finite history can be transformed into an everlasting beauty. Meanwhile, the ever-expanding divine beauty becomes the ultimate context for the ongoing becoming of the world. Out of the infinite divine resourcefulness new definition is continually added

---

14. Whitehead, *Process and Reality*, 346.
15. Ibid. God "saves the world as it passes into the immediacy of his own experience."
16. See chapter 9 for a fuller discussion of the fact of perishing.

to what has already been. And in the divine futurity the entire world becomes forever new, even if many of its temporal epochs are now over.

CONCLUSION

But is such a proposal believable? Certainty, of course, is impossible here. Yet, as Teilhard de Chardin observes, our uncertainty is itself completely consistent with the fact that we and our religions are also part of an unfinished universe. We cannot reasonably expect a theology of nature to answer with climactic clarity the truly big questions humans ask, at least as long as the universe itself is *in via*—and we along with it. Faith reads the universe now only "through a glass darkly," and the darkness and risk that go with faith are somehow inseparable from the fact that the cosmos is still incomplete. However, the incompleteness of the cosmos, my first point, is inseparable from the second—that out of nothingness a world rich in beauty and consciousness has already begun to awaken. If the cosmos is an unfinished story, it is also a story that at least up until now has been open to interesting and surprising outcomes. For fourteen billion years, the universe has shown itself to possess a fathomless reserve of creativity. It has not only been winning the war against nothingness, but in its emergent beauty and its capacity for feeling, thought, and love perhaps it has already begun to taste victory. If the uncertainty in our faith has something to do with the fact that we live in an unfinished universe, then the creative resourcefulness embedded in the same universe cannot fail to give us even now "a reason for our hope" (1 Pet 3:15).

SUGGESTIONS FOR FURTHER READING AND STUDY

Barbour, Ian G. *Religion and Science: Historical and Contemporary Issues.* San Francisco: HarperSanFrancisco, 1997.

Macquarrie, John. *The Humility of God.* Philadelphia: Westminster, 1978.

Morowitz, Harold J. *The Emergence of Everything: How the World Became Complex.* New York: Oxford University Press, 2002.

Pannenberg, Wolfhart. *Toward a Theology of Nature: Essays on Science and Faith.* Edited by Ted Peters. Louisville: Westminster John Knox, 1993.

5

Teilhard de Chardin and the Promise of Nature

The grandeur of the river is revealed not at its source but at its estuary.
—Pierre Teilhard de Chardin[1]

DURING THE PAST CENTURY probably no one has promoted the conversation between science and Christian faith more fervently and effectively than the Jesuit geologist Pierre Teilhard de Chardin (1881-1955). Although, professionally speaking, Teilhard was not a systematic theologian, his reflections on the meaning of the universe from a Christian point of view certainly open up lines of thought that a formal theology of nature cannot overlook.[2] I am devoting this chapter, therefore, to a summary of several of Teilhard's potential contributions to the contemporary dialogue of science and theology.

From his earliest days Teilhard had a deep affection for the natural world, and his wide-eyed perceptivity carried over later to a career in geology and paleontology. He was especially fascinated by rocks, since these symbolized the permanence he sought in the midst of life's perishability. His entire life was a constant search for something incontestably solid to which he might fix his natively anxious spiritual sensibilities. Although he confessed to being temperamentally inclined to tie his life to the siren security of the past, he eventually realized that the world dissolves into the incoherence of mere fragments the farther back we look in time. As his life went on he became increasingly convinced that whatever consistency the universe has lies in its future, not its past. The solidity for which he longed his whole life gradually shifted

---

1. Pierre Teilhard de Chardin, *Hymn of the Universe,* trans. Gerald Vann, O.P. (New York: Harper Colophon, 1969), 77.

2. Teilhard is known foremost as a scientist and not a theologian, but he was familiar with and deeply informed by theologians in the Christian tradition. David Grumett has recently shown in great detail how many strands of theological (and philosophical) tradition went into the shaping of Teilhard's thought. *Teilhard de Chardin: Theology, Humanity and Cosmos,* Studies in Philosophical Theology 29 (Leuven: Peeters, 2005). In my opinion Teilhard produced, among other things, what amounts already to an informal theology of nature.

from the granular multiplicity of the cosmic past—the realm of sheer materiality, in other words—to the unity and power of the cosmic future, where the evolving world would be united climactically to its Maker. It was toward the horizon of the "up ahead" that he looked for the world's foundation. The world, he eventually came to see, "rests on the future . . . as its sole support."[3] It is especially in this respect that his religious instincts merge with the theme of God's promise and futurity, which I am making central in this book's theological reflections on nature.

TEILHARD'S CAREER

Ordained a priest in 1911, Teilhard became a stretcher bearer during the First World War. While his valor in battle was earning him entry into the Legion of Honour, he was becoming increasingly disillusioned with the conventional naturalistic belief that the cosmos can best be understood by looking back to the material simplicity of its causal past, as he thought science was inclined to do. Taking that line of inquiry to its logical conclusion leads eventually to the disarray of primordial cosmic dust. Even science should realize that coherence, and hence nature's intelligibility, emerges only because from its earliest stages the universe has been drawn toward future states wherein emergent complexity gathers atoms, molecules, and cells into the complex unity of organisms. Likewise, the meaning we seek in the cosmos at present, both scientifically and theologically, can reside fully only in the future, not the past or present:

> Like a river which, as you trace it back to its source, gradually diminishes till in the end it is lost altogether in the mud from which it springs, so existence becomes attenuated and finally vanishes away when we try to divide it up more and more minutely in space or—what comes to the same—to drive it further and further back in time. The grandeur of the river is revealed not at its source but at its estuary.[4]

Emboldened by his vision of the cosmic future, Teilhard sought desperately to share with his fellow humans what he thought he could see up ahead. He wrote prolifically, but most of his writings were censored by his superiors and their publication forbidden during his lifetime. His best known book, *The*

---

3. Pierre Teilhard de Chardin, *Activation of Energy*, trans. René Hague (New York: Harcourt Brace Jovanovich, 1970), 239.

4. Teilhard de Chardin, *Hymn of the Universe*, 77.

Phenomenon of Man, appeared in print only after his death. (In 1999 it received a new and improved English translation as *The Human Phenomenon*[5]). Granted that no great writings can escape the historical limitations of their original formulations, some important works nonetheless attain the status of classics to which subsequent ages recurrently turn for nourishment. I believe, along with many others, that Teilhard's *Phenomenon* merits such acclaim. Teilhard was as fallible a thinker as any other great revolutionary in the history of human inquiry, and theology today may not be able to accept every aspect of his work. But because of his nuanced understanding of the relationship of faith to evolution, any Christian theological discussion of nature and science cannot afford to overlook his contributions. It is true that Teilhard draws less attention than he did thirty or forty years ago, but his thought is by no means obsolete. As Jean Lacouture has commented, "The Catholic Church is in great need of the abrasive, energizing breath of a new Teilhard. Or in the interim (why not?) a return to Teilhard? Or, quite simply, a welcome for Teilhard?"[6]

During his career Teilhard worked for many years as a geologist in China—from the 1920s until after World War II. It was during his Chinese sojourn that he wrote not only the *Phenomenon* but also numerous smaller pieces on evolution and faith. This work remained largely unknown to all but a few friends and acquaintances. Returning to France in 1946, he was offered a prestigious academic position at the Collège de France, and once again his superiors refused permission. Following this disappointment, he traveled to the United States, where he found employment at the Wenner-Gren Foundation of Anthropological Research. He participated in two more paleontological expeditions and died virtually alone and unknown in New York City on Easter Sunday, 1955. Hardly noticed during his lifetime, this modest and brilliant Jesuit scientist has turned out to be one of the most important Christian thinkers in the modern period. To those who believe that Christianity—for the sake of its intellectual integrity—must come to grips with science, Teilhard will continue to be an essential resource.[7]

Why were his writings suppressed? Apparently, at least in the eyes of his

---

5. Pierre Teilhard de Chardin, *The Human Phenomenon,* trans. Sarah Appleton-Weber (1959; Portland, Ore.: Sussex Academic Press, 1999).

6. Jean Lacouture, *Jesuits: A Multibiography* (London: Harvill, 1996), 441. The citation is from Grumett, *Teilhard,* 273.

7. To readers interested in pursuing Teilhard's ideas, my own recommendation is to begin with collections of his essays, especially *The Future of Man,* trans. Norman Denny (New York: Harper & Row, 1964); and *Human Energy,* trans. J. M. Cohen (New York: Harcourt Brace Jovanovich, 1962), rather than plunging immediately into *The Human Phenomenon.*

ecclesiastical censors, it was because Teilhard's ideas on evolution and Christianity required too radical a reinterpretation of doctrine. In the aftermath of Vatican I's declaration of papal infallibility and the later condemnations of modernism, a defensive atmosphere had settled into the church's self-understanding at the very time when Teilhard was developing his new vision of Christianity and the cosmos. Church officials feared that the idea of evolution, along with many other innovations in the world of thought, could end up destabilizing Christian doctrine, thereby confusing the faithful. Apprehensions about evolution, not only in the Vatican but also among theologians in general, were especially pronounced in Teilhard's lifetime. Consequently, his enthusiastic embrace of geology, biology, and especially paleoanthropology seemed dangerous, not least because evolution required a new understanding of original sin. Even today, in spite of Pope John Paul II's positive statement on evolution in 1996,[8] many if not most Christians, including theologians, are still reluctant to look very closely and consistently at the possible implications of evolution for Christian faith. It is still difficult to find Christian theologies of nature that are as comfortable with the Darwinian revolution as Teilhard was.[9]

TEILHARD'S VISION

In his religious interpretation of life Teilhard's starting point is a deeply orthodox trust in the Christian doctrines of creation, incarnation, and redemption. The context of his reflections is a scientifically informed understanding of nature quite different from that of early and traditional Christianity. Yet, not unlike many other great Christian writers, he tried to make sense of Christian teaching in terms of the intellectual currency of his own historical period. At the very least, he thought, such an interpretation in our own time requires facing up to evolution. A frank encounter with evolution would be a shock to the religious sensibilities of many people, Teilhard realized, but it is one that honest people must undertake nonetheless.

---

8. John Paul II, "Address to the Pontifical Academy of Sciences" (October 22, 1996), *Origins, CNS Documentary Service* (December 5, 1996).

9. Commentators sometimes misleadingly interpret Teilhard as anti-Darwinian. However, it was the materialist ideology adopted by many Darwinians, not the incontestable empirical data that support evolutionary theory, that he challenged. Like Darwin, Teilhard allowed plenty of room for the role of chance and natural selection, but he rightly objected to the naturalistic belief, which is stronger today than in his own lifetime, that evolutionary mechanisms can provide an *ultimate* explanation of living phenomena.

In my opinion there is no better place for Christians to commence this adventure than by entering through the portals of Teilhard's fascinating way of *seeing* the universe. At first one may find certain aspects of his vision shocking, doctrinally curious, or at least in need of clarification, whether scientifically or theologically. But a Christian theology of nature can learn not only important truths but also courage and honesty from Teilhard's refusal to fortify faith against science. It is worth noting that Teilhard was no more dangerous, bold, or innovative in his own time than Justin, Ireneaus of Lyons, Gregory of Nyssa, and Thomas Aquinas were in theirs.[10] If daringness of vision were an impediment to good theology, this would disqualify many thinkers whose perceived excesses have served to shape Christian tradition.

In any case, Teilhard's expert grasp of natural history convinced him of the need for a radical reinterpretation of Christian teachings about God, Christ, creation, incarnation, redemption, and eschatology in light of the world's and life's ongoing evolution. He became convinced that evolution is not a stumbling block to Christian faith but the most appropriate framework available for clarifying its meaning. As Teilhard saw it, this clarification will not contradict Christian faith, as many fear. Instead it will serve to bring out into the open—with more depth and breadth than ever before—the revelatory image of God's love and futurity.

Whatever one's other criticisms of Teilhard's thought may be, there is no question that his writings have made it possible for many scientifically educated people to remain Christian. This is partly because his ideas arouse hope for the universe and life, thus differing substantially from the acosmic and stylishly pessimistic literature of his own day as well as ours. Theologian Ernst Benz expresses what many readers of Teilhard have felt:

> [Teilhard's] main importance lies in the fact that he opened again the dimension of hope for our time. The opening of the theological aspect of the theory of evolution occurred at a time when the world, or at least the European and American world, got tired of existentialism and theological dialectics. This turning back to an analysis of one's existence, this scorpion-like contortion of the poisonous sting against oneself, this flirting with evil, this digging in the unfathomable depths of one's being, has led to a petrifaction of thinking.[11]

---

10. Grumett, *Teilhard*, 269.

11. Ernst Benz, *Evolution and Christian Hope: Man's Concept of the Future from the Early Fathers to Teilhard de Chardin*, trans. Heinz G. Frank (Garden City, N.Y.: Doubleday Anchor Books, 1966), 226.

Teilhard provides an alternative to such world-weariness. Benz reminds his readers that a whole generation of thinkers and writers who became famous right after World War II were similar to Lot's wife. Like her they "could not look away from the picture of decline and destruction." They "became mesmerized by the abyss of human aberrations" and "got lost in the constricting numbness of fear and defeat . . ." Thought during this period had "turned to stone."[12] But, writing in 1965, Benz remarks that his contemporaries were growing tired of cosmic pessimism and were now paying attention to "thinkers who open their hearts for the beauty of the world and humanity."[13] In 1962, theologian Jean Daniélou had expressed this resurgent mood of hope: "One of the great diseases of the modern mind," he wrote, "is the 'enjoyment' of misfortune, the '*goût du malheur.*' Teilhard detests this with all his heart. And he is right. I wish it were possible to eliminate forever these poisonous miasmas of a decadent Western intelligentsia!"[14]

As biblical scholars and theologians now realize, Christian thought, especially in the West, had over the course of centuries lost touch with the biblical temper of hope. The passionate eschatological expectation that shaped Jesus' religious vision and the thought of the earliest Christian communities had become clouded over, prior to the Second Vatican Council, by centuries of other-worldly spirituality and the corresponding assumption that nothing of lasting significance could possibly be happening in the physical world itself. The eyes of the faithful had been looking upward toward the "next world," while "this world" was thought of as a place to prepare for the soul's journey to paradise. What happened on earth and in natural history, therefore, had no permanent importance other than providing a space in which humans could get ready for heaven. Even today countless Christians assume that the future of the universe matters very little, if at all. Many still doubt that persons are fully part of the natural world or that their destiny is inseparable from that of the whole universe.

Modern biblical studies have demonstrated, however, that the prophets, Jesus, St. Paul, and even the apocalyptic New Testament writings did not think of human destiny as a radical break with the physical universe. Generally speaking, early Christian expectation was turned toward the coming of a *new age* that would transform or re-create the world, not provide an avenue of escape from it. Accordingly, it was the advent of God from out of the future that the early Christian communities awaited, in an intensified version of the

---

12. Ibid.
13. Ibid., 227.
14. Cited by Benz, *Evolution and Christian Hope*, 226-27.

anticipatory spirit of Abraham, Moses, and the prophets.[15] Mostly as a consequence of the Platonizing of Christianity, however, the incarnational emphasis of biblical hope had given way to an increasingly dualistic and acosmic brand of expectation. In Teilhard's lifetime biblical scholars and theologians were still embarrassed by the eschatological enthusiasm of Jesus and the early Christians. They had rediscovered the intense hope of Jesus, the Synoptics, and Paul, but they did not know quite what to make of it. Prominent Christian thinkers even proposed that eschatology be thoroughly demythologized so that God would be thought of more as an eternal present than a destabilizing future.[16] Teilhard, on the other hand, can be credited with a vision that tries to keep alive the biblical openness to a radically new future for the *whole* of reality. He saw in evolution a fresh opportunity to link the entire story of life as well as the history and destiny of the universe to the biblical anticipation of new creation. For this reason alone a theology of nature must take his writings seriously.

TEILHARD AS SCIENTIST

Before looking more closely at Teilhard's theological contributions, it is necessary to say a word about his scientific credentials. During his lifetime Teilhard was held in the highest esteem as a scientist and was commonly acknowledged by his peers to be one of the top geologists of the Asian continent. But since the time of his death and the publication of *The Phenomenon*, his writings have not always met with approval by scientific readers. This distaste is due in great measure to the fact that Teilhard vehemently opposes the predominantly materialist disposition of many prominent scientists, and especially evolutionary biologists. Scientific naturalists are ill-disposed toward Teilhard not so much for scientific reasons as for his conviction that evolution must be carefully distinguished from materialist philosophy.

Evolutionary science makes better sense, Teilhard thought, in a nonmaterialist metaphysical setting, one that gives priority to the future rather than the past. The data of evolution become less, not more, intelligible when interpreted mechanistically or materialistically, that is, by explaining the life story exclusively in terms of the earlier-and-simpler physical causes lying in the dead past. Only by looking toward the estuary, not the source, of the evolu-

---

15. See Jürgen Moltmann, *Theology of Hope: On the Ground and the Implications of a Christian Eschatology*, trans. James W. Leitch (New York: Harper & Row, 1967).

16. Ibid., 106-12.

tionary river will life and the universe begin to disclose their true meaning. As I see it, Teilhard was seeking to replace the materialist "metaphysics of the past" with a "metaphysics of the future," a worldview more in keeping with the Abrahamic and early Christian intuition that ultimate reality comes into the present as an ever-renewing future. Teilhard thought that such a view of reality is essential in order to highlight the fact that evolution is bringing new being, or "more" being, into actual existence.[17]

Much of the negative press Teilhard has received from his scientific critics, therefore, has less to do with disagreement about his scientific research than with his espousal of a worldview opposed to modern scientific materialism. For example, the Nobel laureate and biochemist Jacques Monod, an avowed mechanist and materialist, cast scorn on Teilhard for refusing to go along with what Monod took to be science's decisive expulsion of purpose from nature.[18] Similarly, the late Harvard paleontologist Stephen Jay Gould, annoyed by Teilhard's conviction that there is an overarching directionality to evolution, gratuitously linked him to a notorious scientific hoax.[19] More recently the highly respected American philosopher Daniel Dennett has labeled Teilhard a "loser" simply because he does not go along with Dennett's overt materialism and his claim that evolution entails atheism.[20] The biologist Julian Huxley and the renowned geneticist Theodosius Dobzhansky were enthusiastic supporters of Teilhard as a scientist and visionary, but the American evolutionist G. G. Simpson, in spite of his enjoyment of Teilhard's friendship, would certainly have considered the Jesuit scientist to be deluded in attributing theological meaning to a natural process that evolutionary science had manifestly exposed as purposeless.[21]

Teilhard's professional scientific papers have never been controversial, and most scientists today would still find them impressive.[22] It is Teilhard's better-

---

17. Pierre Teilhard de Chardin, *How I Believe,* trans. René Hague (New York: Harper & Row, 1969), 42-44.

18. Jacques Monod, *Chance and Necessity: An Essay on the Natural Philosophy of Modern Biology,* trans. Austryn Wainhouse (New York: Knopf, 1971), 32.

19. See Stephen Jay Gould's essays in *Natural History,* March 1979; August 1980; and June 1981. For a refutation of Gould's charges against Teilhard, see Thomas M. King, S.J., "Teilhard and Piltdown," in *Teilhard and the Unity of Knowledge,* ed. Thomas M. King, S.J., and James Salmon, S.J. (New York: Paulist, 1983), 159-69.

20. Daniel C. Dennett, *Darwin's Dangerous Idea: Evolution and the Meaning of Life* (New York: Simon & Schuster, 1995), 320.

21. George Gaylord Simpson, *The Meaning of Evolution,* rev. ed. (New York: Bantam Books, 1971), 314-15: "Man is the result of a purposeless and natural process that did not have him in mind. He was not planned . . . Man plans and has purposes. Plan, purpose, goal, all absent in evolution to this point, enter with the coming of man and are inherent in the new evolution which is confined to him."

22. For an authentication of Teilhard's standing as a scientist, see Bernard Towers, *Concerning Teilhard, and Other Writings on Science and Religion* (London: Collins, 1969).

known philosophical and religious reflections on evolution, such as those in the *Phenomenon*, that have raised objections from his fellow scientists. Nevertheless, Harold Morowitz, a widely respected contemporary biochemist, gives a very positive scientific assessment of the *Phenomenon*,[23] even though he does not subscribe to Teilhard's religious vision. Unlike Monod, Dennett, and Gould, Morowitz is fair enough to distinguish clearly between Teilhard as a scientist and Teilhard as a religious thinker. In fact, Morowitz views the *Phenomenon* as contributing much to scientific understanding. He correctly observes, for example, that many years prior to the theory of "punctuated equilibrium" formulated by Gould and Niles Eldredge, Teilhard had already devised an equivalent way to account for the paucity of transitional forms in the fossil record.[24] Morowitz also wonders why Gould, a reputable Harvard paleontologist, would have savagely attacked Teilhard by tying him without any evidence to the notorious Piltdown hoax.[25] Teilhard's scientific critics should also acknowledge more frankly his early anticipation and debunking of the kind of anti-Darwinism so visible today among creationists and proponents of "intelligent design." For Teilhard the reversion to a pre-Darwinian idea of a divine magician performing acts of special creation diminishes instead of honoring the creative power of God.

Teilhard also clearly understood the difference between science and philosophical reflection on science. Unfortunately, however, he did not always make this distinction explicit enough. In the opening pages of the *Phenomenon* he claimed that his book should be read not "as a metaphysical work, still less as some kind of theological essay, but solely and exclusively as a scientific study."[26] Yet the *Phenomenon* is much more than science, at least as science is ordinarily understood. As Ian Barbour has rightly proposed, the fairest way to read most of Teilhard's main works, books such as the *Phenomenon*, *The Future of Man*, and *Human Energy*, is to acknowledge that they are essentially forays into the theology of nature and not simply scientific tracts.[27]

In any case, Teilhard objected to any philosophy or theology of nature that leaves out what is right in front of our eyes, especially the human phenomenon. Unfortunately, when scientists bother at all to investigate the latter, they

---

23. Harold J. Morowitz, *The Kindly Dr. Guillotin and Other Essays on Science and Life* (Washington, D.C.: Counterpoint, 1997), 21-27.

24. Ibid. See also Teilhard's unfortunately neglected work *The Vision of the Past*, trans. J. M. Cohen (New York, Harper & Row,1966). Reading this important set of essays would help remove many of the caricatures of Teilhard as unscientific. The book also includes brilliant defenses of evolution against the attacks of creationists and other critics of Darwinian biology.

25. Morowitz, *Kindly Dr. Guillotin*, 21-27.

26. Teilhard de Chardin, *The Human Phenomenon*, 1.

27. Ian G. Barbour, "Five Ways to Read Teilhard," *The Teilhard Review* 3 (1968): 3-20.

almost always resort to explanatory categories that are unable to illuminate what is distinctively human. They typically bracket out what each of us already knows, "from the inside," to be our most distinctive trait, namely, our subjectivity, our capacity for feeling, thought, decision, and love. Thus, they invariably end up trying to explain the human phenomenon, including the mind, in terms that are too small to clarify its distinctiveness and novelty in evolution. Science, because of its habit of leaving out any mention of the insideness of nature, fails to tell us everything we need to know about what is really going on in the universe.

TEILHARD'S UNIVERSE: WHAT DID HE "SEE"

Teilhard was one of the first scientists in the twentieth century to realize that the entire universe, not just the life story, has a historical character. On our planet, he proposed, natural processes have successively brought about the realm of matter (the *geosphere*), then life (the *biosphere*) and most recently the sphere of mind, or the *noosphere*. The noosphere is the "thinking layer" of earth history, a network made up of human persons, societies, and cultural and technological achievements. Teilhard complained that scientists have failed to *see* that the noosphere is one of the most interesting developments in the history of the universe. Even though the recent emergence of the noosphere is clearly a part of *cosmic* history, ironically it has not yet become a focal topic for cosmology or even earth history. A geologist is conditioned to look for emerging *levels* in planetary evolution, and surely the noosphere is one of these. Yet most geologists, along with cosmologists, have failed to view it as a new layer in natural history continuous with the entire becoming of the universe. Behind this caution lies a scarcely suppressed dualistic suspicion that the world of thought, the human world, is not really part of the universe after all. The phenomenon of thought to this day remains off the map of natural science and most philosophy of nature.[28]

As far as a theology of nature is concerned, it means a great deal that the world of the subject and the phenomenon of "thought" are seen as part of nature rather than remaining essentially alien to it. Teilhard's wider empiricism, which restores the domain of thought to its proper home in nature, places in doubt scientific naturalism's materialist metaphysics underlying the modern assumption that the universe is devoid of purpose. At the same time,

---

28. See B. Alan Wallace, *The Taboo of Subjectivity: Toward a New Science of Consciousness* (New York: Oxford University Press, 2000).

Teilhard's refusal to separate subjectivity or thought from nature as a whole provides theology with a way to make sense of the Christian belief that God acts in nature in a most intimate and effective, even if always mysterious, way.

How so? To begin with, by insisting that the recent emergence of thought in evolution is tied seamlessly into the whole of cosmic history, Teilhard's rich empiricism rules out any dualistic severance of mind or spirit from the physical universe. An essentially mindless realm of matter has never actually existed, since matter was already pregnant with mind (and spirit) from the very beginning of the universe. Once we concede this much, it is no longer unthinkable theologically that the Spirit of God can interact powerfully with the whole physical world. Since the "stuff" of the universe has never been essentially mindless or spiritless, it is permeable to an immanent divine presence in a way that is impossible to conceptualize as long as the universe is thought of along the lines of mechanics alone.

Matter and spirit, in Teilhard's cosmology, are labels for two polar *tendencies* in nature's evolution, not two separate types of substance. "Matter" is the tendency of nature to fall back toward a state of sheer multiplicity and incoherence. "Spirit" is the tendency of nature to move toward unity up ahead. Moreover, it is spirit, not matter, that gives solidity and consistency to the cosmos. Ultimately it is God the Creator who is the center or Supercenter that initiates and grounds the world's inclination toward future unity. It is by unifying things *ab ante*, from "up ahead," that God creates the universe. Since science has informed us that the universe is a still-unfolding process, we may think of nature as suspended between two different kinds of attraction, one toward fragmentation (the past), the other toward communion (the future). So theology should no longer think of the world as divided into separate domains, matter and spirit. Before science discovered that the universe is still coming into being, it was easier to indulge in such partitioning; but since we now realize that the cosmos has always been in the process of becoming, and of giving birth to consciousness, we can more easily understand it as always open to the action of God, even in those earliest phases that we normally consider purely material.[29]

Because its consistency or concrete solidity can be apprehended only inasmuch as matter is becoming spirit, "the cosmos could not possibly be explained as a dust of unconscious elements, on which life, for some incom-

---

29. Pierre Teilhard de Chardin, *The Divine Milieu: An Essay on the Interior Life* (New York: Harper & Row, 1960), 105-11; Pierre Teilhard de Chardin, *Human Energy*, trans. J. M. Cohen (New York: Harvest Books/Harcourt Brace Jovanovich), 57: "In a concrete sense there is not matter and spirit. All that exists [in the created world] is matter becoming spirit."

prehensible reason, burst into flower—as an accident or as a mould."The universe

> is *fundamentally and primarily* living, and in its complete history is ultimately nothing but an immense psychic exercise. From this point of view man is nothing but the point of emergence in nature, at which this deep cosmic evolution culminates and declares itself. From this point onwards man ceases to be a spark fallen by chance on earth and coming from another place. He is the flame of a general fermentation of the universe which breaks out suddenly on earth.[30]

Since there has always been a cosmic tendency toward the emergence of mind and spirit, the cosmos has never been essentially mindless or spiritless.

Why is all of this of interest to a theology of nature? Partly because the intellectual backbone of the modern challenge to Christian faith, and indeed to religions in general, is scientific materialism, a philosophical belief system that claims that the universe is initially, essentially, and finally mindless and spiritless. It is especially this view of reality that so many modern thinkers have made the foundation of their repudiation of Christian faith. Teilhard, therefore, is intent on exposing the shallowness of the materialist creed. He attributes it to a logical fallacy that he calls the "analytical illusion."This is the tendency to view nature "under too great a magnification."The analytical illusion supposes that we can get to the bottom of the phenomena of life and mind by mentally decomposing them into lifeless droplets of "matter."[31] However, concretely there is no such *thing* as matter, and when materialists think they are giving a fundamental explanation for life and mind by pointing to constituent material elements, they are fallaciously mistaking abstractions for concrete entities. Scientific materialism is the consequence of a failure not only of vision but also of logic.

Materialism has no place for either "thought" or the influence of God in nature. But it should be clear by now that our own consciousness is fully part of nature, and that the emergence of thought is the blossoming of a potentiality that has been latent in matter from the very beginning of the universe. This means that there never could have been any moment in natural history when the stuff of the universe was closed off to mind, spirit, or God. Divine action in the world might be hard to envisage if matter were essentially mindless, but the idea of mindless matter is the product of logical illusion and a

---

30. Teilhard de Chardin, *Human Energy*, 23.
31. Teilhard de Chardin, *Activation of Energy*, 139.

failure to "see" nature's insideness. And it is just this insideness that allows a supreme, superpersonal Subject—the Spirit of God, in other words—to interact intimately with nature.[32] It is also the insideness of matter that allows the incarnate and now risen Christ to gather the entire universe, really and not just figuratively, into the crowning majesty of his eucharistic body.[33]

Teilhard, as opposed to the scientific materialists, is convinced that the existence of consciousness—and the universe out of which it has emerged—requires, in addition to the usual scientific accounts, a deeper kind of explanation. And it is his pursuit of this explanatory depth that has rendered his thought so puzzling and often unacceptable to other scientists. Biology and other sciences, Teilhard agrees, can arrive at a more or less adequate account of the outward process of nature's complexification. But what needs fuller explanation is the obvious fact that the outward complexity is accompanied by an increasing inwardness that in humans would eventually become thought.

Inwardness, including reflective self-awareness, is clearly part of the natural world. Thus, it seems deeply ironic that scientific naturalism has systematically turned its focus away from this most obvious and *real* phenomenon of all. Instead of looking at the universe in light of the fact that it has become conscious, scientific naturalism pretends that the universe is essentially unconscious. Then it tries—not without the aura of sorcery—to "explain" the emergent fact of mind solely in terms of what it takes to be an antecedent and underlying mindlessness. To Teilhard this is like trying to pull a rabbit out of a hat. In order to make the impossible project of explaining mind in terms of mindless stuff seem feasible, some scientists and philosophers these days have even gone to the extreme of denying that conscious subjectivity has any reality to it at all. However, Teilhard insists that a deep and wide empiricism cannot plausibly leave the fact of thought, let alone spirit, off any inclusive map of nature.[34]

Thought and spirit are too luminous to be captured by a science accustomed to looking only at the mindless material abstractions that stem from the analytical illusion. To make room for mind in nature, therefore, anyone who professes to see things as they really are must view the pre-human evolutionary world in explanatory terms large enough to allow for the eventual

32. "If the cosmos were basically material, it would be physically incapable of containing man. Therefore, we may conclude (and this is the first step) that it is in its inner being made *of spiritual stuff*" (Teilhard de Chardin, *Human Energy*, 119-20).

33. Pierre Teilhard de Chardin, "The Mass on the World," in Thomas M. King, *Teilhard's Mass: Approaches to "The Mass on the World,"* (New York: Paulist, 2005), 145-58.

34. This point is developed most fully throughout Teilhard's *Human Phenomenon*.

emergence of human subjectivity from within the bowels of nature itself. Modern thought has not been open to such an inclusion. Its picture of the natural world has typically been one in which there is no room for subjects of any sort, let alone ourselves. Teilhard, however, locates human and other subjects in complete continuity with the still emerging cosmos. As far as a theology of nature is concerned this renders the whole cosmos fully transparent to the creative action of God's Spirit, even as it gives cosmic extension to the human subject.

A NEW SPIRITUALITY

It is into this pervasive orientation of the cosmos toward spirit that the incarnate God has descended. The God enfleshed in Christ is the longed-for solidity, the fully concrete and coherent future upon which the universe leans "as its sole support." Theologically speaking, what is *really* going on in evolution is that God is becoming increasingly incarnate in the world (a point witnessed to especially by christological teachings), and the world is exploding "upward into God."[35] Beneath the surface of what science has discovered, there is the eternal drama of God's descent and promise.

Everything that we can make out in the cosmic past with the help of science suggests that the universe is habitually open to further increase in being and value. And there is no reason to think that this openness to *becoming more* has now been suspended or terminated. Theologically speaking, in other words, the universe is still being created and the body of Christ is still in the process of formation. Every eucharistic celebration is a declaration that "what is really going on" even now is that Christ continues to gather us—our labors, joys, and diminishments—along with the entire cosmos into his own body.[36]

The intensifying of consciousness, unity, diversity, beauty, freedom, and love—all of which Teilhard refers to as the birth of "spirit"—is the supremely important thing going on in the universe, all of it culminating in the world's redemption in Christ. Thus, it would be a pity for Christians to go through life without an awareness that we are all being invited at each present moment to involve ourselves in the great work of increasing the *universe's* own being, that is, of making the cosmos, as the extended body of Christ, *more* than what it is and has been. But without first *seeing* that something momentous is already going on in the universe, our hopes and moral aspiration will lack the

---

35. Teilhard de Chardin, *Future of Man*, 83.
36. Teilhard de Chardin, "The Mass on the World."

power, drama, and spirit of adventure they might otherwise have. This is why science is so important for Christian spirituality.

I believe with Teilhard, therefore, that the greatest shortcoming of contemporary spirituality, Christology, and also of most ethical theory, whether religious or secular, is to have missed out on the universe. By failing to see that our own lives are part of a cosmic stream that flows, however haltingly, toward the ocean of what is more, our moral life lacks an adequately energizing incentive and begins to drift aimlessly, stoically, and anthropocentrically on a silent sea. In his "Mass on the World," Teilhard prays, "Shatter, my God, through the daring of your revelation the childishly timid outlook that can conceive of nothing greater or more vital in the world than the pitiable perfection of our human organism."[37] These words should help dispel the often repeated but erroneous charge that Teilhard's thought is noxiously anthropocentric. It is a *cosmic* transfiguration of hope and virtue that he calls for, and he wants us to feel deeply that we are all part of a cosmic drama that our religious ancestors knew nothing about.

Perhaps these days we might also adopt such a sweeping perspective as we look for a unifying basis for a global and ecologically responsible ethic. However, so far the cosmos has been virtually absent in otherwise noble attempts to arrive at a planetary moral consensus. It still seems that most ethicists overlook the fact that our earth is part of a much larger universe in which something unimaginably immense, beautiful, mysterious, and momentous is working itself out. Ethicists and spiritual directors seldom consider the prospect that human action could be redirected and invigorated by an awareness that our planet has an important part to play in a much larger, indeed cosmic and Christic, drama of creation.

The general assumption in most modern and contemporary spirituality is that what is going on in the universe has very little to do with what is going on terrestrially, ecologically, nationally, and ecclesiastically, as well as in our personal and family lives. However, the human phenomenon, as Teilhard emphasizes, is a "function of sidereal evolution of the globe, which is itself a function of total cosmic evolution."[38] To overlook the cosmic roots of obligation, he thinks, is to leave ourselves intellectually, ethically, and spiritually stranded.

Teilhard would wonder how many of us consistently take into account in our thoughts about God and Christ the cosmic context of our lives. Doing so would in no way entail the forsaking of time-tested religious doctrines or

---

37. Ibid., 150.
38. Teilhard de Chardin, *Human Energy*, 22.

virtues such as humility, gratitude, moderation, justice, and love. Rather, these would now have fresh meaning. In a more cosmic setting, ethical activity and worship of Christ will mean that our faith, hope, and love are participating in the ongoing creation of the *universe*. For Christian spirituality and ethics this means connecting our search for the Kingdom of God or our building the body of Christ to the ongoing creation of the heavens and the earth. A cosmically reformed spirituality does not imply that we would stop doing the small and often excruciatingly monotonous things we have always had to do. But at least we may now situate even the most uninteresting duties and obligations of everyday life within the framework of hope for an entire universe's fulfillment. Such a perspective, Teilhard believes, could ennoble and energize our modest efforts.

Teilhard was deeply troubled by the persistent religious dualism that continues to separate the moral responsiveness of religious people from the universe itself. This severance only "sickens" Christianity, he says, preventing it from feeling the enlivening sap that rises up from cosmic roots. A "zest for living" must be the underpinning of all serious ethical endeavor, but such vitality requires the conviction that our efforts have the backing of the universe.[39] By failing to recognize that we are now being invited to participate in the great work of cosmic creation, religious obligation—or, for the Christian, the following of Christ—tends to become a matter of obeying arbitrary categorical imperatives or extrinsic divine commands, seeking a reward in the hereafter, or at best improving ourselves and the human condition—but for what? Cut off from an awareness of our being part of an imposing cosmic drama, and lacking a full appreciation of the doctrines of creation and incarnation, Christian ethical life becomes a matter of "killing time," and redemption a mere "harvesting of souls" from the universe.

For healthy and robust hope to exist, there must always be room for something *more* to happen, since nothing "clips the wings of hope" or subverts the incentive to enthusiastic moral action more severely than the religious assumption that everything great and important has already occurred in some splendid mythic or cosmic past, and that the most human effort can lead to therefore is a restoration of what once was.[40] Only a "passion *for being finally and permanently more*," an opening toward the future of creation, says Teilhard, can sustain and invigorate our spiritual and ethical lives.[41]

---

39. Teilhard de Chardin, *Activation of Energy*, 231-43.

40. Pierre Teilhard de Chardin, *Christianity and Evolution*, trans. René Hague (New York: Harcourt Brace Jovanovich, 1969), 79.

41. Teilhard de Chardin, *How I Believe*, 42.

MORAL EFFORT AND THE IMPERMANENCE
OF THE UNIVERSE

Finally, however, for Teilhard the incentive to do good also requires a trust that our efforts can have a *lasting* impact on the whole of things. And since the universe itself will eventually perish, Teilhard observes, there must be a more permanent guarantee that our spiritual aspirations and moral efforts are not ultimately futile. As an informed scientist, Teilhard does not deny that entropy predicts a temporal termination of the physical universe. But the prospect of a "total death," he adds, would "immediately dry up in us the springs from which our efforts are drawn . . ."[42] From a Christian point of view, apart from the gathering of all events into the body of Christ, and hence into the eternal life of God, it would be unreasonable to trust that the long cosmic story of creation would add up to anything at all in the end.

To many of our contemporaries the whole idea of such a preservative divine Care is a blind stab in the dark, and I shall take up a discussion of this skepticism in chapter 9. Let it suffice for now to say that a widely empirical awareness of what is going on in the universe, a way of seeing that takes us beyond the analytical illusion, can hardly fail to notice the perpetual emergent miracle of an anticipatory universe that for fourteen billion years has in fact been bringing more being into existence out of less. So for Teilhard it is not an irrational leap to trust that the same future horizon—he calls it "God-Omega"—that so far has faithfully kept the universe open to the emergence of "what is *more*," can finally, in the compassionate embrace of the cosmic Christ and in the fellowship of the Holy Spirit, take the whole of evolution and the effects of human endeavor permanently into its everlasting and healing embrace as well. Even then—as I think Teilhard would agree—there would always and forever be ample room for still *more* to occur.

SUGGESTIONS FOR FURTHER READING AND STUDY

Grumett, David. *Teilhard de Chardin: Theology, Humanity and Cosmos.* Studies in Philosophical Theology 29. Leuven and Dudley, Mass.: Peeters, 2005.

Teilhard de Chardin, Pierre. *Christianity and Evolution.* Translated by René Hague. New York: Harcourt Brace Jovanovich, 1969.

———. *The Future of Man.* Translated by Norman Denny. New York: Harper & Row, 1964.

———. *Hymn of the Universe.* Translated by Gerald Vann. New York: Harper Colophon, 1969.

---

42. Ibid., 43-44.

6

Evolution and Divine Providence

Unlike Teilhard de Chardin, the majority of Christians have had a very difficult time coming to grips with the new biology. The main issue, now as always, is how to reconcile evolution with the idea of divine providence.[1] After Darwin, what does it mean to say that God "provides" or cares for the world?[2] Prior to the age of science, of course, suffering and evil had always made people ask whether God truly cares for the world, but the relatively new awareness of evolution has made that question more relevant than ever. What does "divine providence" mean if life comes about and diversifies on earth in the manner proposed by Darwin and his scientific descendants?

Prior to Darwin, belief in divine providence matched comfortably the static hierarchical worldview that had prevailed for centuries. The dominant world picture in Western thought after Plato and Aristotle generally consisted of a "Great Chain of Being," running from lowly matter at the bottom to divine creative wisdom at the top.[3] The intervals between matter and God were occupied by plants, animals, humans, and angelic beings, all contributing to a rich vertical plenitude of creation wherein every being had its divinely assigned station. Moreover, the differentiation of levels allowed humans to feel radically distinct from other living beings. Created "in the image and likeness of God," followers of biblical faith could assume they were cared for in a special way by God on high.

However, let us compare this venerable cosmography with the scientific understanding of the universe as pictured in the introduction to this book.

---

1. This chapter draws some material from my Boyle Lecture "Darwin, Design and the Promise of Nature," given in London in February 2004 and published in *Science and Christian Faith* 17 (2005): 5-20; and my Sophia Lecture at Washington Theological Union (published as "What If Theology Took Evolution Seriously," *New Theology Review* 18 [November 2005]: 10-20).

2. In responding to this question, I shall deliberately overlook the traditional distinction between general and special providence, especially since the former often tends to think of God in a deistic manner while the latter often does so in an excessively interventionist sense.

3. Arthur O. Lovejoy, *The Great Chain of Being: A Study of the History of an Idea* (New York: Harper & Row, 1965).

Recall the thirty large volumes, each 450 pages long, and every page standing for one million years in a roughly fourteen-billion-year-old universe. Life waits until volume 22 to make its debut, about 3.8 billion years ago; the Cambrian explosion, with its rapid diversification of life forms, does not occur until the last pages of volume 29; and our hominid ancestors start to appear only several pages from the end of volume 30. "Modern humans" endowed with a capacity for intelligence, ethics, religious aspiration, and scientific inquiry show up only in the last several lines on the last page of the final book.

Our question here, then, is whether it will be possible to rewind the vine of providential care that had clung for centuries to the hierarchical cosmology around the new thirty-volume horizontal narrative of a world still in process of emerging? Can the religious sensibilities and aspirations fashioned in the context of ancient and medieval cosmologies survive at all, let alone find new life, in the new picture of an evolving universe?[4]

Even prior to Darwin it was possible to picture a gradual unfolding of the Great Chain of Being over a long period of time. The principle of plenitude, which holds that every level of the hierarchy must be occupied by an appropriate set of beings, could in principle be implemented by a providential "program" for a more leisurely unfolding of creation.[5] In other words, belief in divine governance is not in principle incompatible with an evolutionary understanding of nature. However, the specifically Darwinian recipe for life's emergence and proliferation over the last four billion years seems to many Christians and religiously skeptical biologists to be incompatible with any God who really cares for the world. In the words of the philosopher David Hull, biological evolution is "rife with happenstance, contingency, incredible waste, death, pain and horror." Any God who would allow life to evolve in the manner depicted by Darwin's science is "careless, indifferent, almost diabolical." This is not, Hull adds, "the sort of God to whom anyone would be inclined to pray."[6]

Darwin's understanding of evolution consists of three generic features that seem, at least at first sight, to suggest anything but the gentle guidance of divine providence. In the first place, evolution entails *accidents* in abundance, and these seem antithetical to any divine plan. Even the origin of life now

---

4. For further discussion, see my book *God after Darwin: A Theology of Evolution* (Boulder, Colo: Westview, 2000).

5. Charles Coulston Gillispie, *Genesis and Geology: A Study in the Relations of Scientific Thought, Natural Theology, and Social Opinion in Great Britain, 1790-1850* (Cambridge, Mass.: Harvard University Press, 1996), 18.

6. David Hull, "The God of the Galapagos," *Nature* 352 (August 8, 1992): 486.

strikes scientists as purely spontaneous rather than vigilantly planned by a Creator. Additionally, the genetic variations (mutations) that provide the raw material of evolutionary change and diversity are purely random—and that means undirected. Accidents in the larger sweep of natural history also shape the story of life. Climate changes, ice ages, earthquakes, volcanic eruptions, asteroid impacts, and the like have twisted the trails of evolution in the most unpredictable ways. For example, about sixty-five million years ago it seems that a large object from outer space crashed into the earth with enormous force. The impact produced an explosion of such magnitude that the earth's atmosphere was drastically altered, large reptiles such as the dinosaurs became extinct, and their departure ushered in the age of mammals. The eventual evolution of primates, hominids, and humans seems to have been dependent on an accidental catastrophe in natural history that few of us would have called providential if we had witnessed it live.[7] How then can we reconcile devout trust in God's care with the high degree of accident, or contingency, in evolution?

Second, Darwin's recipe includes most notably *natural selection*, the merciless mechanism that eliminates all nonadaptive forms of life. Natural selection separates the "fit" from the "unfit" without regard for their intrinsic value, feelings, or urge to live.[8] A major question for theology, today no less than in the years immediately subsequent to the publication of Darwin's *Origin of Species*, is how to reconcile divine providence with the blindness, struggle, suffering, death, and waste characteristic of natural selection.[9]

Scientists are now realizing, it is true, that there is more to evolution than the brutality that natural selection brings along with it. Evolution also requires cooperation within and among species. Nevertheless, in any future revisions of the theory it is not likely that the idea of natural selection will be discarded, even though the precise extent of its creative role in evolution may be forever debated. In any case, theology will always be obliged to deal with evolution's apparent impersonality.[10] In the long story of life, natural selection has left behind it a deep gorge of loss and pain. Consequently, many sensitive

---

7. Even if one accepts the theory of evolutionary "convergence," according to which unrelated phyla can develop similar features such as eyes or wings, this in no way destroys the fact that contingency is abundant in the life story. See Simon Conway Morris, *Life's Solution: Inevitable Humans in a Lonely Universe* (New York: Cambridge University Press, 2003).

8. Biologically speaking, "fitness" means the probability an organism has of reproducing, that is, of getting its genes into the next generation.

9. See Gillispie, *Genesis and Geology*, 220.

10. For a discussion of some previous attempts in England to connect God and evolution, see Peter Bowler, *Reconciling Science and Religion: The Debate in Early Twentieth-Century Britain* (Chicago: University of Chicago Press, 2001).

people have given up trying to make any religious sense of it. A theology of nature, on the other hand, is obliged to face the implications of natural selection head-on.

Third, in order to be richly creative, the combination of accidental variations and blind selection requires an enormous amount of time. In fact, the universe as understood by contemporary cosmology has given life a span of several billion years for experimentation. The new awareness of how much time it has taken to get from microorganisms to humans also seems to challenge religious trust in divine providence. The evolutionist and outspoken atheist Richard Dawkins proposes that if life had had only a biblical span of six thousand years or so, this would not be enough time for primitive cells to evolve into something as complex as the eye or the human brain. In such a relatively short period of time the intervention of special divine creativity would be required, and so it would make sense to believe that life is "intelligently designed" or at least nudged along by divine providence. But what if the journey of life on earth, as measured by contemporary science, has had as much as 3.8 billion years to advance from the earliest instances of metabolism to human brains? In that amount of time small, successive, accidental changes in organisms, only a tiny proportion of which are preserved by natural selection, can eventually produce improbable outcomes, such as the eye or brain, without any need for divine involvement. According to Dawkins and many other contemporary evolutionary naturalists, Darwinism and deep time have made it unnecessary to appeal to divine action or divine providence in order to explain the amazing complexity and diversity of life.[11]

THE TASK OF THEOLOGY AFTER DARWIN

A theology of nature must ask, therefore, whether the three main ingredients of Darwin's recipe (accident, natural selection, and deep time) are compatible with the doctrine of providence. It cannot overlook the inordinate amount of suffering and loss that evolution permits. The question of life's suffering is such an important matter that I shall take it up more explicitly later in this chapter. For the present I want only to note that the idea of divine providence, which has generally been associated closely with a divine "plan," "purpose," or "design," does not seem to follow closely the Darwinian charting of life's journey. In Western religious thought the careful cataloging of evidence for divine

---

11. This is the central argument of Richard Dawkins in *Climbing Mount Improbable* (New York: W. W. Norton, 1996).

design was traditionally one of the main tasks of "natural theology," but after Darwin this venerable practice seems futile to many theologians as well as scientists.

In 1802 William Paley's work *Natural Theology,* a work that Darwin himself later read with youthful admiration,[12] had argued that if the intricate design of a watch points to the existence of an intelligent watchmaker, then the even more elaborate "contrivances" in nature, such as the eye or heart, entail the existence of an intelligent and benign Creator. But in the wake of Darwin's *Origin of Species* (1859), many Christian theologians have gradually abandoned arguments for God's existence based on the complex design exhibited by living organisms. Today it is mostly creationists and proponents of intelligent design who promote this facet of natural theology, but in doing so they reject the carefully gathered evidence for evolution from geology, comparative anatomy, embryology, biogeography, radiometric dating, and especially genetics.[13]

Most Christian theologians informed about evolution would not wish to defend the claim that the complex design in life can lead the mind directly to God. They might agree with Cardinal John Henry Newman, who even before Darwin's world-shaking publications had stated that he had little use, religiously speaking, for Paley's natural theology. He thought Paley's approach could "not tell us one word about Christianity proper," and "cannot be Christian, in any true sense, at all." Paley's brand of "physical" theology, Newman goes on to say "tends, if it occupies the mind, to dispose it against Christianity."[14]

Today, in keeping with scientific method's focus on purely physical causes, evolutionary biologists explain the intricate design of organisms without any

---

12. William Paley, *Natural Theology; or, Evidence of the Existence and Attributes of the Deity Collected from the Appearances of Nature* (Edinburgh, 1816; ed. with a new introduction by Matthew D. Eddy and David Knight, Oxford: Oxford University Press, 2006).

13. Advocacy of intelligent design appears in Phillip E. Johnson, *The Wedge of Truth: Splitting the Foundations of Naturalism* (Downers Grove, Ill.: InterVarsity, 1999); Jonathan Wells, *Icons of Evolution: Science or Myth? Why Much of What We Teach about Evolution Is Wrong* (Washington, D.C.: Regnery, 2000); Michael J. Behe, *Darwin's Black Box: The Biochemical Challenge to Evolution* (New York: Free Press, 1996); William A. Dembski, *Intelligent Design: The Bridge between Science and Theology* (Downers Grove, Ill.: InterVarsity, 1999). For critiques of intelligent design theory, see John F. Haught, *Deeper than Darwin: The Prospects for Religion in the Age of Evolution* (Boulder, Colo.: Westview, 2003); and Kenneth R. Miller, *Finding Darwin's God: A Scientist's Search for Common Ground between God and Evolution* (New York: Cliff Street Books, 1999).

14. John Henry Newman, *The Idea of a University* (1854; Garden City, N.Y.: Image Books, 1959), 411. Newman is also famous for saying that he believes in design because of God, not in God because of design. See the discussion of Newman by Alister McGrath, "A Blast from the Past? The Boyle Lectures and Natural Theology," *Science and Christian Belief* 17 (April 2005): 29-30.

appeal to divine engineering. Adaptive design—for example that a fish's eye, unlike that of land mammals, is able to see clearly under water—appears almost miraculous. But, as far as science is concerned, the citing of mystery and miracle here must give way to purely natural explanations. To scientifically uninformed people the fish's eye may *seem* to be the product of intentional design by God, but to a scientist it is enough to say that the round eye of a fish is the outcome of accident, natural selection, and deep time, the three ingredients in Darwin's recipe. As evolutionists tell the story, any marine animals in the remote past that may have had, say, oval-shaped rather than round eyes could not see well enough to detect predators, so they were eaten up before having the opportunity to survive and reproduce. Meanwhile, those that by accident had been endowed with a potential to develop round-shaped eyes turned out to be more "fit." They were better equipped to see under water and detect predators, and so they had a higher probability of surviving and reproducing. They passed on genes for round eyes to subsequent generations, and that helps explain why underwater animals can adapt to an aqueous environment today. Similar stories can be told about the origins of all species and the many features of organisms without ever invoking the ideas of mystery and miracle.

Does divine providence or wisdom, therefore, have any role whatsoever to play if the emergence of living diversity is so natural a process? To naturalists, adaptive complexity has been brought about by a combination of accident, selection, and deep time *rather than* by divine governance. Prior to Darwin, belief in God seemed to mesh smoothly with natural theology's sense of life's design. But now evolutionary biology explains all the characteristics of living organisms without recourse to divine agency.[15] Natural selection of minute random variations over a long period of time seems sufficient to account for every instance of organic complexity as well as of organisms' behavioral tendencies. The enormous amount of time that radiometric dating has now made available to scientific speculation has given much more confidence to evolutionary biologists than Darwin himself could enjoy. Today he would be reassured by the generous outlay of 3.8 billion years that evolution has had available for its many experiments, most of which have been unsuccessful. Consequently, the "evidence of design" that gave rise to earlier versions of natural theology seems to have dissolved in the acid of Darwinism. What justification, then, could there possibly be for adhering to the doctrine of divine providence amid the darkness of evolution?

---

15. Gary Cziko, *Without Miracles: Universal Selection Theory and the Second Darwinian Revolution* (Cambridge, Mass.: MIT Press, 1995.)

PURE TRUST?

One response is to make an unconditional act of faith in God's providence and wisdom in spite of evolution's unwieldy way of creating life's diversity.[16] The fact that devout people cannot make theological sense of the Darwinian recipe should not prevent them from trusting blindly that there is nonetheless a hidden meaning in evolution, one to which it would be presumptuous on our part to try to gain access. Maybe, therefore, "contingent," "accidental," "random," "wasteful," and "purposeless" are terms that we ignorant mortals attach to evolution only because of our abysmal ignorance of God's wider vision for the universe. Whenever anything takes place that falls outside our habitual ideas of decent design, we tend to refer to it as an accident or absurdity. But perhaps it is *really* part of a wider and eminently wise divine plan to which we are not privy—and into which we should not pry.

Perhaps Darwin's demolition of design, though a defeat for natural theology, is a victory for faith. For all we know, what appears in evolution to be absurd contingency from a rationalist perspective could be the tangled underside of a tapestry that, from God's vantage point on the other side, is a tightly ordered pattern. As for the struggle, cruelty, waste, and pain in evolution—evidence, at least to many, of a universe beyond the pale of providence—Darwin's recipe has absolutely nothing qualitatively new to add to the perennial challenges to pure faith. True piety, by its very nature, is already aware of the realities of evil and suffering, but it trusts *in spite of* all apparent absurdity. Indeed, too much intellectual self-confidence may even deaden a faith that can become fully alive only where there is radical uncertainty.

According to this fideistic (faith-alone) approach, our contemporary awareness of life's evolutionary struggle may pose no more of an obstacle to trust in divine providence than fate, suffering, death, and evil have always done. There is no attempt here to reject evolutionary biology, and indeed one can be a faith-alone Christian and at the same time an excellent biologist.[17] But there can never be a rational reconciliation of Christian faith with the ruthlessness of Darwin's recipe. Consequently, since human (including scientific) perspectives are always limited, we are not in a position to state with certainty, as evolu-

---

16. For example, Robert E. Pollack, *The Faith of Biology and the Biology of Faith* (New York: Columbia University Press, 2000).

17. If I read him correctly, Simon Conway Morris, an evolutionary paleobiologist at Cambridge University, a devout Christian and rightly celebrated scientist, seems, at least at times, to represent the fideistic approach. See his reaction to my own approach in "Response to the Boyle Lecture," *Science and Christian Belief* 17 (April 2005): 21-24.

tionary naturalists do, that the Darwinian universe is at bottom indifferent or malicious. Science is not opposed to faith, though we cannot say why.

PROVIDENCE AS PEDAGOGY?

Such blind faith, however, can at best tolerate evolution. It can never celebrate it or integrate it tightly into a theological vision. One intriguing attempt to justify evolution theologically is based on the ancient religious intuition that earth was set up by God as a school for life, and as a pedagogical setting designed especially for the making of souls. The idea is that life and souls can grow only in the presence of challenges. After all, how anemic of character would humans be if the terrain over which their own lives traveled were utterly devoid of obstacles and dead ends? If earth were a totally accommodating abode, our lives would languish to the point of pure passivity. Perhaps the Darwinian curricular regime was set up deliberately by providence to educate human beings for their eventual graduation to eternal life.[18]

Once our species emerged, the Darwinian syllabus that had already given rise to so much living abundance and diversity could now function as the pedagogical context for molding human character as well. What more efficient environment for soul-making could any of us have conjured up than the one that Darwin described a century and a half ago? Pursuing this question, the science writer Guy Murchie asks his readers what kind of a world they would have created if they were God. Would it be one in which people could all luxuriate in blissful tranquility and undisturbed hedonism? Or would it not be a world approximately like the one that Darwin has given us?

> Honestly now, if you were God, could you possibly dream up any more educational, contrasty, thrilling, beautiful, tantalizing world than Earth to develop spirit in? . . . Would you, in other words, try to make the world nice and safe—or would you let it be provocative, dangerous and exciting? In actual fact, if it ever came to that, I'm sure you would find it impossible to make a better world than God has already created.[19]

The Darwinian world, therefore, is the best of all possible worlds after all, and divine providence is manifest in the rigor and ruthlessness, not just the creativity, of evolution.

---

18. See the comparable suggestions by John Hick, *Evil and the God of Love*, rev. ed. (New York: Harper & Row, 1978), 255-61, 318-36.

19. Guy Murchie, *The Seven Mysteries of Life: An Exploration in Science and Philosophy* (Boston: Houghton Mifflin, 1978), 621-22.

The idea that God is a practitioner of tough love can be found scattered throughout the Bible, especially in Hebrews 12:5-13. It is also espoused by many notable theologians in Christianity's history. But it is not hard to imagine how such an outlook on suffering could at times build resentment and even hatred of God for making the world so unnecessarily severe. Moreover, a soul-making theodicy becomes so narrow at times that it portrays the entirety of creation as serving primarily the goal of *human* salvation. This picture of providence renders the story of the universe inconsequential except as a stage for putting souls through a series of trials to prepare them for salvation. Nowadays this evolution-as-curriculum proposal seems excessively anthropocentric and too earth-centered for refined ecozoic and cosmic tastes.[20] Even if suffering were a proper punishment for our sins, we would still have cause to wonder, along with Darwin, why there is such a surfeit of pain in the lives of other species of sentient beings. And why do they have to go to "school" with us in the first place?

TOWARD A THEOLOGY OF EVOLUTION

In contrast to the proposals just summarized, a Christian theology in touch with science will ask whether the processive character of the cosmos overall, and the troubling Darwinian recipe in particular, is incompatible with what one should expect if the world is grounded in and providentially embraced by the God revealed in Jesus Christ. It will reflect on evolution more in the light of Christianity's revolutionary understanding of God than in terms of the "intelligent designer" of antievolutionist religious thought. If Jesus is the visible face of the invisible God, the one in whom the fullness of God dwells (Col 1:19), "the radiance of God's glory and the exact representation of his being" (Heb 1:3), then a theology of evolution would do well to make this teaching its point of departure. It will ask whether the revelatory image that gives rise to the Christian understanding of God (that of God's descent and promise) is also sufficient to bring intelligibility to the life world as evolutionary science now understands it.

This task may not be easy. For if nature by itself is able to bring about living diversity in what appears to evolutionists to be a blind and unguided manner, where would there be room for theological illumination of the process?

---

20. This, I believe, is a major failing of John Hick's soul-making theodicy as expressed in *Evil and the God of Love,* in spite of several attempts on his part to acknowledge the dangers of anthropocentrism.

Furthermore, since Christians believe that God cares for the weak and the poor, why is the process of natural selection permitted to eliminate so unfeelingly all incompetent, maladaptive forms of life? And if Christians believe that God is eager to create goodness and redeem all suffering, why is the creative process drawn out so inconclusively and for so many billions of years?

In view of these difficulties, theology may effectively display the harmoniousness of divine providence with Darwinian evolution if it takes as its starting point the two standout features of the revelatory image of God that I have been using to focus this book's inquiries. The first of these, as we have seen, is that of the "descent of God."[21] The second, a motif that permeates the biblical literature, is that of the God who opens up the future by making promises and dependably keeping them. I believe that on these two closely allied pillars of faith Christian theology may at least begin to construct a plausible theology of evolution.

The Descent of God

In Philippians 2:5-11 Paul pictures Jesus as being "in the form of God." But, not clinging to that status, Jesus empties himself and takes on the form of a slave. Theological reflection has often taken this understanding of Jesus by nascent Christianity (possibly derived from an early liturgical hymn) to imply that what is really being emptied out is the very being of God. But the notion of a divine emptying (*kenōsis*) does not depend solely on a single text from Philippians. It is Jesus' entire life and death on the cross that reveal to Christians the astounding notion that God is essentially humble, self-emptying love that gives itself away unreservedly to the entirety of creation for the sake of divinizing that creation. In fact, in the light of God's self-revelation in Christ, one may read the entire body of biblical narrative as straining to tell the story of the humble descent of the infinite God into the domain of creaturehood—all for the sake of deeper intimacy with and elevation of the world.

As theologian Donald Dawe writes, "God accepted the limitations of human life, its suffering and death, but in doing this, he had not ceased being God. God the Creator had chosen to live as a creature. God, who in his eternity stood forever beyond the limitations of human life, had fully accepted these limitations. The Creator had come under the power of his creation. This

---

21. For a useful study of the history of this theme, see Joseph M. Hallman, *The Descent of God: Divine Suffering in History and Theology* (Minneapolis: Fortress, 1991).

the Christian faith has declared in various ways from its beginning." However, Dawe continues,

> the audacity of this belief in the divine kenosis has often been lost by long familiarity with it. The familiar phrases "he emptied himself [*heauton ekenosen*], taking the form of a servant," "though he was rich, yet for your sake he became poor," have come to seem commonplace. Yet this belief in the divine self-emptying epitomizes the radically new message of Christian faith about God and his relation to man.[22]

A theology of nature, I would add, will endorse this kenotic perspective as it applies not only to God's relationship to persons but also to the *entirety* of creation.

Theologian John Macquarrie, echoing Dawe, notes how radical the transformation of the God image has been in Christianity:

> That God should come into history, that he should come in humility, helplessness and poverty—this contradicted everything . . . that people had believed about the gods. It was the end of the power of deities, the Marduks, the Jupiters . . . yes, and even of Yahweh, to the extent that he had been misconstrued on the same model. The life that began in a cave ended on the cross, and there was the final conflict between power and love, the idols and the true God, false religion and true religion.[23]

An evolutionary theology, I would suggest, may picture God's descent as entering into the deepest layers of the evolutionary process, embracing and suffering along with the *entire* cosmic story, not just the recent human chapters. Through the liberating power of the Spirit, God's compassion extends across the totality of time and space, enfolding and finally healing not only human suffering but also all the epochs of evolutionary travail that preceded, and were indispensable to, our own emergence.

In spite of its endless diversity, there is a fundamental unity to the life process. All of life is linked, throughout its evolution, to the trinitarian life of God. Redemption must mean, then, that the whole story of the universe and life is embraced by divine providence. Not simply gathering our human sto-

---

22. Donald Dawe, *The Form of a Servant: A Historical Analysis of the Kenotic Motif* (Philadelphia: Westminster, 1963); see also Jürgen Moltmann, *The Crucified God: The Cross of Christ as the Foundation and Criticism of Christian Theology*, trans. R. A. Wilson and John Bowden (New York: Harper & Row, 1974); and Hans Urs von Balthasar, *Mysterium Paschale: The Mystery of Easter*, trans. by Aidan Nichols, O.P. (Edinburgh: T & T Clark, 1990).

23. John Macquarrie, *The Humility of God* (Philadelphia: Westminster, 1978), 34.

ries into the trinitarian drama, the God revealed in Christ assimilates also the whole story of life on earth (and life elsewhere in the universe if it exists there). And since the entire physical history of the universe, as recent astrophysics has made clear, is tied into the existence of life everywhere, theology can no longer separate human hope for ultimate deliverance from the larger cosmic course of events. Because of the divine omnipresence, nothing in the universe story or in life's evolution can occur outside of God's own experience. If Jesus is truly the incarnation of God, then his experience of the cross is God's own suffering. And, by virtue of life's unbroken historical unity, Christian theology may be so bold as to assume also that the eons of evolutionary suffering in the universe are also God's own suffering. This would mean that the whole of nature in some way participates in the promise of resurrection as well.

But how does the theme of the divine descent help us understand why life, in the first place, has unfolded in the frayed manner that evolutionary science depicts? Why are struggle and death constitutive of the ongoing creation of life? And once life comes about, why does God's creation and providential governance not follow the path of pure design that natural theologians and intelligent design proponents would prefer?

There is great mystery here, and at this point theology becomes highly speculative. But once again a kenotic theology of creation may be enlightening.[24] For if theology remains true to its revelatory sources, it must also envisage the divine descent as the ground of creation itself. That is, even as a condition of there being any world distinct from God at all, the omnipotent and omnipresent Creator must be humble and self-effacing enough to allow for both the *existence* of something other than God, and an ongoing *relationship* to that other. If the creation is to be truly other than God, and not just an accessory attached to God's own being, then the divine omnipotence and omnipresence would become "small" enough to allow room for what is truly distinct from God—although it must be added that this self-constraint is paradoxically a function of God's greatness. It is out of the infinite largesse of the divine humility, therefore, that the otherness of the world is "longed" into being by God. Creation is God's "letting be" of the world, a release that makes possible a dialogical relationship (and hence an intimate communion) of God with the finite, created "other."

Once God's "other," a world fashioned in accordance with the eternal Logos, emerges as a historical reality, it can sustain its otherness only by

---

24. See, e.g., Jürgen Moltmann, *God in Creation: A New Theology of Creation and the Spirit of God*, trans. Margaret Kohl (San Francisco: Harper & Row, 1985), 88.

becoming more, not less, differentiated from, its Creator.[25] And once life emerges spontaneously within the historical unfolding of the world, it need not forfeit the autonomy accorded to it by the other-regarding goodness of its maker. Perhaps, then, it is *ultimately* because of God's self-abandoning humility that the Darwinian recipe consists of its three ingredients: contingency, lawful constraint, and abundant time.

Contingency, for instance, may be troubling to those fixated on the need for design in nature, but an openness to accidents seems essential for creation's autonomy and eventual aliveness. The alternative would be a world so stiffened by lawful necessity that everything in it would be eternally dead. A world devoid of accidents would not really be a world at all, but a pointless puppet.[26] At the same time, the seemingly impersonal laws of nature, including natural selection (the second ingredient of the Darwinian recipe), may be theologically understandable if the world is to have any degree of consistency, autonomy, or self-reliance vis-à-vis its Creator. A lawless universe would be mere chaos. Finally, if nature is allowed to be distinct from the God who calls it into being, it must be granted sufficient time for life-evolving experiments to take place within the context of the wide variety of possibilities made available to it by the infinite resourcefulness of its Creator. Once we understand divine providence as inseparable from the eternal divine descent, we may appreciate why the creation of the world does not take place instantaneously but instead unfolds across billions of years.

The God revealed in Jesus allows the world enough scope to become ever more distinct from—and more intensely related to—its creative ground. The principle enunciated often by Teilhard that "true union differentiates" applies both to the internal life of the Trinity and also to the God–world relationship. True union is not homogeneity or uniformity but a relationship that paradoxically allows the conjoined entities to realize a deeper freedom and distinctiveness than they could find in isolation.[27] A mature theology of providence may come to realize, therefore, that God "descends" from all eternity not to absorb the world or overwhelm it with the divine presence, but instead to ennoble it by supporting its self-actualizing. Thus, evolution is com-

25. As Teilhard often notes, differentiation is an effect of, not a deterrent to, true union. See, e.g., *The Human Phenomenon*, trans. Sarah Appleton-Weber (1959; Portland, Ore.: Sussex Academic Press, 1999), 186-87.

26. Even St. Thomas Aquinas found the idea of a world without accidents or contingencies theologically inconceivable; see *Summa contra Gentiles*, III, 74. See Christopher Mooney, S.J., *Theology and Scientific Knowledge* (Notre Dame and London: University of Notre Dame Press, 1996), 162.

27. Pierre Teilhard de Chardin, *Activation of Energy*, trans. René Hague (New York: Harcourt Brace Jovanovich, 1970), 116.

pletely consonant with a world of emergent freedom that allows for an ever more intimate encounter with God.

Providence and Promise

The evolution of life on earth has required a nearly unimaginable amount of time, but Darwinian science cannot give an adequate account of time itself. Evolutionary biology neither asks nor answers the question of why the universe has been endowed with an irreversible temporal character at all. Given enough time, evolutionists rightly point out, even the most improbable things can happen. But time must first be given! Why then is the universe temporal at all? The various sciences may respond to this question in their own ways, but a biblically guided theological search for the deepest ground of temporality, and hence of the possibility of evolution, may assume that it is *the coming of the future* that pushes the present into the past and permits a linear sequence of events to occur. In other words, it is not the blind movement of the past toward the future that endows the universe with its temporal character. Rather, it is the constant arrival of a new future.

Theologically speaking, however, the "coming of the future" is ultimately the coming of God, whose self-revelation occurs inseparably from promises that open up the world to an unprecedented horizon of newness. Indeed, it is not inappropriate to suppose that, in its ultimate depths, the fathomless future is one of the things we mean when we use the word "God" in a Christian setting.[28] The arrival of the future in the mode of promise explains in an ultimate way not only the fact and depth of time but also the other two ingredients of evolution's recipe, contingency and lawful predictability. The contingency in natural history, even when accompanied by pain and loss, is essential to any world open to the coming of the future. For if nature were completely devoid of undirected, accidental events it would be so rigid that it could never become new and alive at all. Instead it would persist in endless cycles of deadening sameness.

Likewise, the predictable constraints that we call the laws of nature, including natural selection, are essential to the world's openness to the future.

---

28. See Jürgen Moltmann, *The Experiment Hope*, ed. and trans. M. Douglas Meeks (Philadelphia: Fortress, 1975), 48; Karl Rahner, *Theological Investigations*, vol. 6, trans. Karl and Boniface Kruger (Baltimore: Helicon, 1969), 59-68; Wolfhart Pannenberg, *Faith and Reality*, trans. John Maxwell (Philadelphia: Westminster, 1977), 58-59; Ted Peters, *God—The World's Future: Systematic Theology for a New Era*, 2nd ed. (Minneapolis: Fortress, 2000).

If it were devoid of the recurrent routines operative in its unfolding, the universe could have no narrative continuity from one stage to the next. At every moment it would collapse back into droplets of disarray. Deep down beneath Darwin's evolutionary recipe there resides what may be called the "promise of nature."[29] It is on this foundation—an exquisite narrative blend of contingency, predictability, and temporal openness—that evolution is able to purchase a foothold in nature. Providence, then, consists not of direct, manipulative engineering of cells and organisms, as the proponents of intelligent design teach, but of caring enough for creation to allow it to unfold from within itself as a dramatic, suspenseful, and momentous story of emergent freedom. And the prediction that the Big Bang universe will itself eventually come to a physical ending should not be disturbing to those who trust in God's promise of redemption. After all, an infinitely compassionate and resourceful Future can be the ultimate redemptive repository of the entire series of cosmic moments no less than of those episodes that make up our own personal lives.

THEOLOGY AND THE SUFFERING OF SENTIENT LIFE

It is beneficial for theology to steep itself fully in the Darwinian portrait of life. Taking evolution seriously can lead to richer thoughts not only about God and nature but also about the meaning of suffering. This implies, however, that a theology of nature must pay attention to the suffering of all sentient forms of life, not just that of humans. A daring look at the *whole* story of life, especially the pre-human chapters of evolution, invites Christian theology to give voice to the entire universe's anticipation of the redemption promised by God.

Evolutionary naturalists, however, see no need for any theological comment on life's suffering at all. For them, every aspect of life, including suffering, can be explained sufficiently in biological terms. From the perspective of Darwinian biology, suffering (which I take here to be inclusive of the sensation of pain, fear, and loss by all instances of sentient life) is nothing more than an evolutionary adaptation. The capacity for suffering increases the fitness of complex organisms; that is, it enhances their probability of surviving and reproducing. To make complete sense of suffering there is no need to

---

29. I have developed this theme in my book *The Promise of Nature* (Mahwah, N.J., and New York: Paulist, 1993).

invoke religious ideas at all. Can a theology of nature, therefore, add anything of substance to the evolutionary naturalist's account?

Darwin himself remarked that suffering is "well adapted to make a creature guard against any great or sudden evil."[30] Suffering is adaptive because it warns living beings endowed with nervous systems that they may be in danger. In some strains of contemporary evolutionary thinking a capacity for suffering makes good sense if life, at bottom, is simply a matter of genes finding their way into subsequent generations. In this seductively simple understanding, genes provide instructions that equip organisms with sensitive warning systems. These elaborate systems of communication in turn help keep organisms alive long enough to pass their genes on to subsequent generations. Ultimately it is strands of DNA that govern the whole show, and suffering is simply one way in which genes enhance reproductive fitness. So, what could theology possibly add to the elegance of this evolutionary account of why suffering occurs?

Before addressing this question directly, we need to take note of the fact that DNA also programs organisms to perish. Every living thing eventually dies, and once again evolutionary biology explains why. If organisms never died, new and more complex species of life would not have any room on a small planet to exist and evolve. Natural selection requires an enormous amount of genetic diversity, and a single generation of organisms is not enough. Multiple successive ages of organic proliferation are necessary if evolution is to get anywhere, including to humans; so the deaths of countless ancestral organisms has been an absolute necessity. It all makes good sense biologically.

Yet we cannot help wondering about the startling *excess* of suffering and loss, any more than Darwin could. Why is there so much pain and death in the story of life, especially since a lot less would still be enough to ensure the survival of genes? The sheer volume and intensity of suffering in sentient life throughout the vast epochs of evolutionary time, including especially the very recent period of human and cultural evolution, stagger the sensitive mind. Science has now made it abundantly clear that our earth was never a paradise at any time in the past. Suffering, death, and occasional mass extinctions attended the life story long before humans showed up on page 450 of volume 30 of our thirty-volume cosmic story. Without extinctions, biologists agree, it

---

30. Charles Darwin, *The Autobiography of Charles Darwin, 1809-1882: With Original Omissions Restored*, ed. Nora Barlow (New York: Harcourt, 1958), 88-89.

is unlikely that mammalian, primate, and human life would ever have come about at all. We owe our existence and our organic complexity to the sacrifice of innumerable generations of organisms and now defunct species.

Christian theology, it would seem, should not ignore this protracted chronicle of pain and sacrificial loss that prevailed long before the religious species finally came along. Christians hope that eventually all tears will be wiped away and death will be no more, but does this apply to all of life? Theology is not accustomed to exploring fully and frankly the question of what evolutionary suffering might mean for our understanding of God, creation, redemption, and eschatology.[31]

Evolutionary Naturalism and the Suffering of Sentient Life

The question "why suffering?" is ancient and persistent. It has given rise to fascinating myths about the origins of our own travail and how we might find release from it. But myths and theologies about the origin and end of suffering, including those of Christianity, came to expression long before Darwin. So they had nothing to say about the heartlessness of natural selection and its eventual eradication of most of the species of life that have ever existed on earth. They knew nothing of deep time and the endless struggle of life to exist and survive. But this understandable omission provides no justification for contemporary Christian theology's ignoring the longer evolutionary trail of striving and pain.

Traditionally, Christian thought has attempted to respond to the problem of suffering, and evil in general, with what is known as "theodicy." In its strictest sense a theodicy is any attempt to "justify" the existence of God in the face of evil and suffering. If God is all-good and all-powerful, then God must be able and willing to prevent life's suffering. But suffering and evil clearly exist. Theodicy tries to say why, and in the process it seeks to exculpate the Creator for permitting suffering. Even though theodicies are seldom very helpful, they are nonetheless significant expressions of the perennial human need to make sense of suffering, and so they need to be taken seriously. Furthermore, the term "theodicy" can be stretched, as the sociologist Peter Berger proposes, so that it means *any* attempt to make sense of suffering.[32] Therefore,

---

31. Among recent attempts to take into account the suffering (and pain) of nonhuman sentient life, John Hick's *Evil and the God of Love* is one of the most impressive. Yet even Hick's theodicy is not deeply informed by an evolutionary sense.

32. Following the broad usage of the term by sociologist Peter Berger in *The Sacred Canopy: Elements of a Sociological Theory of Religion* (Garden City, N.Y.: Anchor Books, 1990), 53ff.

even Darwinism can function as a kind of theodicy, since it explains life's capacity for suffering tidily in terms of the notion of evolutionary adaptation.

To evolutionary naturalists, moreover, Darwinism surpasses all previous theodicies in clarity and simplicity. For many thoughtful people today evolutionary explanation makes any theological appeal to such ideas as sin and expiation superfluous as far as the meaning of suffering is concerned.[33] Of course, learning about the ways of evolution will scarcely wipe away all tears or destroy death forever. But to the evolutionary naturalist, being cleansed of religious hope is not too large a price to pay for the crisp simplicity of the Darwinian solution to so intractable a problem as that of suffering.

How, then, may a Christian theology of nature respond to what has become an intellectually attractive evolutionary justification of suffering? Unlike traditional Christian reflection, a theology of nature today must keep its attention firmly fixed on the suffering of pre-human and nonhuman forms of sentient life in evolution. Can the revelatory portrayal of God's descent and promise embolden Christian theology to take into account the wider natural history of evolutionary suffering? If so, such a broadened perspective would require a decisive shift away from theodicies that are still dominated by the theme of expiation. In what follows I can provide little more than a sketch of this proposal.

Christian Theodicy and the Suffering of Sentient Life

What would be the consequences for theology of holding before our eyes the pre-human history of predation, disease, pain, death, and extinction along with what evolutionary science, geology, and cosmology have shown to be the still unfinished state of the universe? It is hard to say, since even our most sophisticated theodicies have yet to undertake, let alone complete, such an assignment.[34] Christian theodicies have traditionally attended almost exclusively to *human* suffering, and they have connected human suffering primarily to guilt and sin. Recently Christian thought has begun to shift away from a strictly expiatory understanding of suffering, but it still provides very little in the way of an extended consideration of evolution's wider motif of misery.[35]

---

33. For a fuller discussion, see my book *Is Nature Enough? Meaning and Truth in the Age of Science* (Cambridge: Cambridge University Press, 2006).

34. See, e.g., John Thiel, *God, Evil, and Innocent Suffering: A Theological Reflection* (New York: Crossroad, 2002).

35. Thiel's otherwise rich discussion of theodicy scarcely mentions evolution, an omission shared by most contemporary Christian theologians.

A major reason for clinging to the classical theological emphasis on human guilt as the cause of suffering is that it easily absolves the Creator of any responsibility for life's anguish. It is not God, but we, who are to blame. However, evolutionary understanding forces us now to envisage suffering and death as an essential part of the larger story of creation. Science has demonstrated that an enormous mound of mortality, disease, and slaughter lies buried beneath the presently flourishing façade of life-forms and human culture. At times theologians have reacted to this relatively new understanding of nature by playing down, or even denying, any significant suffering in non-human organisms.[36] But a sense of evolution brings out the extreme anthropocentrism of such a point of view. So a responsible theodicy must ponder not only human suffering but also the fact that the creative process on our planet seems to have occurred at the price of continual pain and perishing.[37]

In the light of our revelatory image of God's descent, may we envisage the suffering and dying of all of life as, at least in some sense, God's own suffering and dying? In the light of God's promise of new creation may we also have a justifiable hope that the entire unfinished universe will eventually arrive at a redemptive fulfillment? Finally, is it possible that expectation—hope in the promises of God—may come to replace expiation as the dominant Christian response to unjustified suffering?

According to most traditional Christian theodicies, human suffering is justified because of our sinful acts of rebellion against the Creator. God has set things up in such a way that human suffering is necessary to pay the price for transgression. So, only if guilt is appropriately expiated will there come an end to suffering. The religiously sensitive philosopher Paul Ricoeur refers to this kind of theodicy as the "ethical vision" of evil.[38] It is this interpretation of evil and suffering that underlies the "Adamic myth." Here the figure of Adam represents the biblical intuition that an all-beneficent Creator has no complicity in evil and that responsibility for life's suffering must lie in human fault. In our own solidarity with Adam we have all gone astray and hence deserve to

---

36. Thiel (parallel to C. S. Lewis and John Hick) allows that animals may experience "pain" but not suffering. Even making this distinction, however, seems to take the issue of theodicy as pertaining only incidentally to nonhuman life (*God, Evil, and Innocent Suffering*, 1-31).

37. For Whitehead it is the fact of perishing, the vanishing into the past of every present moment, that constitutes the ultimate evil in the transient world. See Alfred North Whitehead, *Process and Reality*, corrected ed., ed. David Ray Griffin and Donald W. Sherburne (New York: Free Press, 1968), 340.

38. Paul Ricoeur, *The Conflict of Interpretations: Essays in Hermeneutics*, ed. Don Ihde (Evanston, Ill.: Northwestern University Press, 1974), 455-67. According to the ethical vision of evil, "punishment only serves to preserve an already established order" (Paul Ricoeur, *History and Truth*, trans. Charles Kelbley [Evanston, Ill.: Northwestern University Press, 1965], 125).

suffer proportionate punishment. Christianity understands Jesus as the one exception to such straying, and because his suffering is completely innocent it can be understood as adequate payment for our own guilt. Consequently it can release us from at least some of the suffering we truly deserve.

Ricoeur points out, however, that the figures of Job and the Suffering Servant represent a theological project, starting centuries before Christ, of shifting theodicy out from under its subordination to the motif of expiation. The innocence of the Servant, the one who bore our guilt and was "pierced for our transgressions," opens the way to understanding suffering as gift rather than sheer punishment (Isa 53:5-6). So the erosion of a strictly expiatory understanding of suffering is already occurring in the biblical traditions prior to Christianity.[39] All the more, then, reflection on evolution in the context of Christian faith may lead contemporary theology to question whether the ethical vision, which has played so prominent a role in religious thought and spirituality, should still remain dominant in theodicy.

If suffering can have the character of gift (or grace) rather than expiation, this can only be because the ultimate subject of life's suffering is nothing less than God, since "love comes from God" (1 John 4:7). In some sense, therefore, Christian theology may now suppose that God's descent takes the suffering of *all of life* into the divine experience. And it may trust accordingly that the power of the future that arrives on the wings of the revelatory promise can eventually destroy the suffering and death that accompany the whole journey of life. In any case, it does not seem wise for theology or theodicy to ignore the wider drama of life's many instances of suffering and self-sacrifice as having nothing to say about God. Instead it should seek to rescue some meaning from it all. Eons of living, subjective centers have given themselves over to suffering, death, and even extinction, all for the sake of allowing new life (including our own) to rise up continually on the ruins of their great self-surrender.

God's identifying with the suffering of Jesus entails solidarity with all human suffering. But why stop here? In view of the evolutionary continuity and kinship that we now know to exist between ourselves and all other kinds of life, it would be unjustifiably arbitrary to overlook God's sharing in the wider-than-human story of life's suffering. Hence a raw encounter of theology with evolution may deepen our sense of God's own descent as well as the extent of God's redemption and divinization of the world. The pervasively *sacrificial* character of innocent life in evolution's creative advance does not have

---

39. This is one of the main arguments of Paul Ricoeur's *The Symbolism of Evil*, trans. Emerson Buchanan (Boston: Beacon, 1969).

to be pictured as taking place outside the trinitarian drama. Christian faith encourages us to see the entirety of evolution as occurring within God's life.

The End of Expiation?

If the struggle of life is God's own struggle, then this would justify theology's moving the evil of pain further than ever from the clutches of a purely expiatory theodicy. The innocence of life's victims means that the theme of sacrifice has to be decoupled radically from the ethical vision's emphasis on suffering as punishment. Where there is no guilt there is no need for retribution.

Of course, in the presence of life's extensive suffering, one still has the option of taking a naturalistic stance and interpreting suffering as pure tragedy that can only be endured courageously. Perhaps the "meaning" of suffering is that it provides individual persons the opportunity to feel new strength in the face of fate and death, in the manner of Albert Camus' Sisyphus. Ever since Darwin science has hammered home the fact that most suffering is innocent, and such a tragic fact subverts any narrowly ethical vision of evil. So to many naturalists the excessive suffering of sentient life is reason enough for embracing a tragic vision of existence. But for Christians, perhaps life's tragic suffering is best understood as creation's calling out for redemption by God (Rom 8).

The expiatory understanding of suffering, however, is so deeply embedded in our sensibilities that it seems nearly ineradicable. It first took verbal shape in ancient stories about how an original cosmic perfection was spoiled by free human acts of rebellion. In the biblical world the Adamic myth represents the intuition that suffering exists mostly because of human sin.[40] An offshoot of this influential theodicy has been that people are inclined to look for culprits even today, in secular as well as religious societies, whenever suffering or misfortune occurs.[41] The assumption that a price in suffering must always be paid for the defilement by human freedom of a primordial purity of creation has underwritten the entrenched habit of looking for victims. It has legitimated a history of scapegoating that has only exacerbated violence and misery.[42]

In 1933 Pierre Teilhard de Chardin wrote:

---

40. Even in the Adamic myth, however, the figure of the serpent stands for the intuition that evil is not reducible simply to human productivity (Ricoeur, *Conflict of Interpretations*, 294-95).

41. Pierre Teilhard de Chardin, *Christianity and Evolution*, trans. René Hague (New York: Harcourt Brace Jovanovich, 1969), 81.

42. Ibid.

In spite of the subtle distinctions of the theologians, it is a matter *of fact* that Christianity has developed under the over-riding impression that all the evil round us was born from an initial transgression. So far as dogma is concerned we are still living in the atmosphere of a universe in which what matters most is reparation and expiation. The vital problem, both for Christ and us, is to get rid of a stain. This accounts for the importance, at least in theory, of the idea of sacrifice, and for the interpretation almost exclusively in terms of purification. It explains, too, the pre-eminence in Christology of the idea of redemption and the shedding of blood.[43]

What Teilhard is questioning, at least by implication, is our habit of associating sacrifice primarily with expiation. Unfortunately, theology still sometimes exaggerates the idea of a hypothesized primordial offense, which in turn usually assumes that God's original creation was one of rounded-off perfection. Especially in Western Christianity theologians have situated suffering in the context of myths that idealize the beginning rather than the future of creation. This only makes the original fault seem all the more enormous, hence unleashing demonizing expeditions to find someone or something to blame. The logic operative here is that if a state of paradisal wholeness had preceded the original fault, then the fault itself could have been no trivial matter. An expiatory view of suffering and sacrifice would then be called upon to help us understand how to make things right. What is worse, setting things right would mean the *restoration* of what has been, rather than an opening to the future of a completely new creation up ahead.

It is important to ask, therefore, just what theological consequences would follow if the universe, as evolution implies, has emerged only gradually from a state of relative simplicity and still remains unfinished. What cosmic need would there be for expiation or scapegoating if nothing significant had been lost at the time of cosmic origins? And what if the perfection for which humans long were envisaged as a future creation instead of a forfeited past? Is it not an imagined sense of the *enormity* of what has been allegedly besmirched by sin that breeds a kind of remorse that looks for restoration via expiation, and that, at its worst, fosters interminable resentment and a spirit of revenge? If the original breach had been less consequential, or indeed if there had been no universe-shattering rebellion against goodness in any mythic or historical past at all, would there still be any need for sacrificial restoration? Would not an evolutionary view of life logically call for a theology that purges sacrifice of its motifs of expiation and situates life's suffering and sacrifice once and for all within the horizon of a redemptive future?

---

43. Ibid.

I can only leave these as questions for future theological reflection. I am asking, to sum things up here, what might be the consequences for theology were it to think out fully and conclusively the implications of the evolutionary claim that *a state of complete cosmic integrity* in the realm of created being has never (yet) been an actuality? My own suspicion is that by ruling out any past epoch of created perfection, our religious aspirations may henceforth be turned decisively away from remorseful nostalgia and refocused in the direction of hope. Christ entered the temple of sacrifice once and for all. The age of expiation, therefore, is now altogether a thing of the past (Heb 10:1-18).[44]

The evolutionary character of nature is difficult to square with a backward-looking nostalgia for a hypothesized state of original perfection.[45] But evolution *is* compatible with hope for a final future fulfillment. That Christianity is essentially a religion of the future should make theology leap with excitement at the fact that evolution is inconsistent with seductive dreams of reinstituting a glorious past perfection. Evolution has closed off this retrograde path to salvation once and for all. In the wake of Darwin and contemporary cosmology, after all, it is difficult for most educated people to believe that there has ever been any point in past cosmic history when the universe was pristinely perfect. Accordingly, there could not have been any literal "fall" from a cosmic paradise into the state of imperfection. Imperfection would have been present from the start, as the shadow side of an unfinished universe.[46] Hence, it would follow that the cosmological substructure of expiatory self-punishment, resentment, and victimization would never have been a fact of nature in the first place. Since there could never have actually taken place in the history of nature the cataclysmic loss of primordial perfection that might justify our resentment at such an imagined bereavement, there would be no justification for persistent acts of expiation or for seeking out culprits to blame for the miseries of our condition. Scapegoating violence would make no sense in a universe that is still coming into being.

Unfortunately, however, the story of human religiosity has often been more one of nostalgia for an imagined past perfection than an eschatological anticipation of new creation. I believe this is still the case. Even in religions descended from the biblical environment a longing to restore or recover an

---

44. Gerd Theissen, *The Open Door: Variations on Biblical Themes*, trans. John Bowden (Minneapolis: Fortress, 1991), 161-67.

45. John Hick, in a manner similar to Friedrich Schleiermacher, tries to save the idea of an original human perfection by redefining perfection to mean having the possibilities for development (Hick, *Evil and the God of Love*, 225-41). But in my view the very definition of perfection is the "full actualizing of possibilities." And from all the evolutionist can see, the world's possibilities are still far from being fully actualized.

46. Teilhard, *Christianity and Evolution*, 40.

idyllic past has worked to suppress the spirit of Abrahamic adventure that turns us toward the uncertain future opened up by a God of promise. What, then, would be the implications of situating the longed-for realm of perfection in the not-yet-future instead of in a remote cosmic past?

The Theodicy Problem Redefined

At the very least it would seem that the task of theodicy henceforth should not be to fit the fact of suffering onto the grid of guilt and punishment. Instead, if it hopes to get closer to the truly substantive issue, theodicy might ask why an all-good and all-powerful God would create an *unfinished, imperfect, evolutionary* universe in the first place rather than one that is complete and perfect from the beginning. Could it be that if creation is to take place at all a truly good and deeply powerful God has no choice but to call the world into being in an evolutionary way? This is material for much theological reflection, but the short answer may be that any imaginable world that is completely finished and perfected *ab initio* would not be distinct from God and could not really be a creation at all. Perhaps an originally finished creation, as Teilhard and others have emphasized, is theologically inconceivable.

In view of the image of God's descent, the theodicy problem discussed above proves to have been wrongly stated. As it is usually understood, the question is how to reconcile the fact of suffering and evil with God's infinite goodness and power. However, the meaning of the term "power" as used in typical formulations of the theodicy problem is precisely what the Christian revelation has challenged. If power means the capacity to manipulate, then it is difficult if not impossible to reconcile the existence of life's suffering with the existence of God. But the Christian revelation consists in great measure of the radically new message that the meaning of power has been forever transformed by its intimate conjunction with self-giving Love. The idea of divine power has not been destroyed, only transfigured. For if power means the capacity to bring about significant effects, then the God of self-emptying love can bring about effects that a manipulative power could never accomplish.

Among these not the least is freedom. Only a God of self-limiting love could allow beings endowed with freedom to exist. Freedom, after all, is not something that can be caused, since to cause something, at least on the model of efficient causation, is to determine it, to situate it in a series of events in which one state follows by necessity from another. Freedom can arise in nature only spontaneously, as something not deterministically fashioned. It can exist and thrive only in the presence of a noncoercive, generous environment of "letting be." Theologically understood, creaturely freedom is a response to an infinite love whose very definition is to let something *other* than itself come into being.

However, in order for human freedom to arise in continuity with natural processes, the whole universe must be thought of as having come into being in the very same environment of selfless love that has permitted human freedom to arise very recently. It is because of the depths of this love that the creation of the universe cannot be coercively completed in the beginning, but unfolds in an experimental, temporally drawn out manner. The universe, at least for now, is unfinished, and hence imperfect, because of God's humble, selfless love of otherness. A specifically Christian theology, therefore, may find considerable solace, liberty, and hope in a universe that is still emerging into being.

Of course, an evolutionary theology of nature will remain entirely faithful to the history of the religious longing for perfection, since human vitality requires pursuit of an ideal. However, it is not necessary to picture the perfection to which our hearts aspire as though it were something that once existed and has now been lost. It may be more appropriate instead to picture perfection as a state that has never yet been actualized but that we may hope will come into being in the future in accordance with God's vision of what is good, true, and beautiful. One of the important implications of evolution for theodicy is that it allows for the transpositioning of the ideal of perfection from an imagined past to a possible future.

The biblical accounts of creation and promise are themselves struggling to bring about just such a radical reconfiguration at the roots of the human longing for perfection. The ancient narratives of a *promising* God, a God who always opens up a new future whenever dead-ends appear, encourage us to move beyond wistful obsession with a lost Eden outward into an open future that transplants the essential domain of perfection into the unimaginably resourceful territory of the "up-ahead," in the direction of a creation yet to be realized. The Bible's eschatological orientation arouses hope for an unprecedented future, even as it deflects our pining for a paradisal past. By our participating in a "great hope held in common"[47] the roots of violence are numbed and human energy cooperatively directed toward the horizon of new being.

The Question of Original Sin

I want to end this chapter by insisting that what I have said here in no way entails a diminishment of a sense of sin, or of the need for genuine remorse

---

47. Pierre Teilhard de Chardin, *The Future of Man*, trans. Norman Denny (New York: Harper & Row, 1964), 75.

for the evil that human beings bring about personally and socially, or for our own urgent need of redemption. In fact, just the opposite is the case. Even the idea of "original sin," emphasized more by Latin than Eastern Christianity, can still make good sense. It now means that each of us is born into and "stained" by a world wherein a long history of human refusal to attend to the Spirit's call for creativity, hope, and love has accumulated, stamping its destructive mark deeply into societies, families, and persons. In evolutionary terms, our species' collective resistance to the call to creativity, hope, and love and its desperate resignation to the entropic pull toward fragmentation—that is, to a metaphysics of the past—have enfeebled persons, society, and nature.

The adaptive environment for life, freedom, and hope is an unfinished universe. But since it is unfinished it is also still imperfect. This means it can also provide a foothold for the existence and perpetuation of evil. And because the fault line of evil extends all the way down into the very depths of the universe, the drama of evil and the need for salvation are in no way diminished but instead indefinitely enlarged by the evolutionary and cosmic setting in which theology henceforth has to dwell. Evolution places in question the dominance of the expiatory vision, but it in no way entails a diminishment of the scope of salvation by Christ. Christ's incarnate, redeeming, and divinizing presence can now be magnified to unprecedented cosmic proportions by the discoveries of natural science. The *unfinished* character of this immense universe means that there is a primordial and persistent need for redemption in the very marrow of cosmic reality, a need that is as wide as the heavens and as long as time itself. It is this universal imperfection as well as our personal and social sins that calls out for redemption, and hence for a Savior that we can truly worship.[48]

SUGGESTIONS FOR FURTHER READING AND STUDY

Edwards, Denis. *The God of Evolution: A Trinitarian Theology.* New York: Paulist, 1999.
Haught, John F. *God after Darwin: A Theology of Evolution.* Boulder, Colo.: Westview, 2000.
Miller, Kenneth R. *Finding Darwin's God: A Scientist's Search for Common Ground between God and Evolution.* New York: Cliff Street Books, 1999.
Teilhard de Chardin, Pierre. *The Human Phenomenon.* 1959. Translated by Sarah Appleton-Weber. Portland, Ore.: Sussex Academic Press, 1999.

---

48. Teilhard de Chardin, *Christianity and Evolution*, 39, 54.

7

Cosmology and Creation

THE IDEA THAT THE UNIVERSE is created is really quite shocking. In the history of thought many notable thinkers have found it quite unbelievable, if it has even occurred to them at all. They have assumed instead that the universe is eternal and uncreated. Democritus, Plato, Aristotle, and most other ancient philosophers would have found the Christian attribution of creation to a personal God exceedingly strange. And, until very recently, most scientific naturalists likewise have taken it for granted that the physical universe has existed always.

To biblical faith, however, the doctrine of creation implies that the universe is neither self-originating nor necessary. The universe is the contingent (as opposed to necessary) product of God's unlimited graciousness. Since the world does not have to exist, therefore, it can be received by us as pure gift. The ultimate foundation of the world's existence is the goodness and power of God. Theologically speaking, Jürgen Moltmann proposes, it is God's descent or self-humbling love that allows something other than God to exist at all.[1] The otherness that occurs within God's trinitarian life—when the Father, in a primordial self-emptying, begets the Son—is what makes possible the creation of the universe in the image of the Logos, the Word of God.[2] In order to be other than God the universe cannot have been finished or perfected immediately in the beginning (for reasons to be discussed later), and so Christian revelation persuades us to view the inestimable gift of the universe as the embodiment of a promise yet to be fulfilled through the divinizing and consummating power of the Holy Spirit. Between the primordial self-giving of God and the final perfecting of all things by the Spirit lies the great adventure of creation.

By maintaining that God creates the world *ex nihilo*, out of nothing, Christian teaching safeguards its belief in the complete liberty of God to allow a

---

1. Jürgen Moltmann, *God in Creation: A New Theology of Creation and the Spirit of God*, trans. Margaret Kohl (San Francisco: Harper & Row, 1985), 88.

2. For an excellent discussion of the Trinity, see Anne Hunt, *Trinity: Nexus of the Mysteries of Christian Faith*, Theology in Global Perspective (Maryknoll, N.Y.: Orbis Books, 2005).

distinct world to exist. Accordingly, the consistent Christian teaching of *creatio ex nihilo* makes ample room for gratitude, the most fundamental of religious responses. If the cosmos existed either by necessity or as a sheer accident—as though it either "had" to be or "just happened" to be—there would be no need to give thanks for it. *Creatio ex nihilo* means, among other things, that the world is a free gift. To deny the freedom of God to create is to render thanksgiving unnecessary, and hope for new creation superfluous. Today it is not unusual, however, for naturalists to claim that modern cosmology renders dispensable the notion of creation by God.[3] In doing so they deliberately disown one of the central beliefs of Christianity, Judaism, and Islam.

To the Abrahamic traditions the doctrine of creation is not a story devised for the sake of satisfying mundane curiosity about how everything began. The philosophy of nature and, more recently, natural science are quite capable of providing that kind of information, although their ideas remain subject to revision. The doctrine of creation, on the other hand, points more profoundly toward the permanent "ground of being" that underlies the passing universe. In the depth of all finite being, as theologian Paul Tillich puts it, there is the everlasting "power of being" that we call God.[4] I have been suggesting that Christian revelation, with its emphasis on the theme of promise, understands the power of being simultaneously as the "power of the future."[5] God, in other words, is also the font of endless renewal. The same power that brings the world into being and sustains it in existence can bring permanence to what has perished and new life to what is now dead. Christian belief in resurrection builds on the doctrine of creation.

CREATION AND THE BIG BANG

The Big Bang theory of the universe embraced by almost all contemporary cosmologists is usually taken to imply that the universe had a beginning, approximately fourteen billion years ago. Science itself has now apparently overturned the idea of an eternal universe. Even the fact that scientists ascribe

3. For example, Peter Atkins, *Creation Revisited* (New York: W. H. Freeman, 1992).

4. Paul Tillich, *Systematic Theology*, 3 vols. (Chicago: University of Chicago Press, 1963), 2:10-12, 20, 125.

5. Wolfhart Pannenberg, *Faith and Reality*, trans. John Maxwell (Philadelphia: Westminster, 1977), 58-59; Ted Peters, *God—The World's Future: Systematic Theology for a New Era*, 2nd ed. (Minneapolis: Fortress, 2000).

a finite number of years to the universe's period of existence implies logically that the cosmos had a beginning. But does the scientific conclusion that the universe had a definite beginning, and hence has not been around forever, add any meaningful support to the doctrine of creation?

Not necessarily. Even if the universe has existed forever, Christian theology must reply, it could still never have been separated from a divine power of being and renewal.[6] Even an eternal universe could still be said to be a created one. The doctrine of creation, in other words, is not so much about how things began as about why there is anything at all rather than nothing. So a theology of creation is in no way dependent for its meaning or plausibility on contemporary cosmology. In fact, it is never prudent for theology to tie itself too closely to any current scientific consensus.[7]

Nevertheless, it is not without interest to theology that science itself, at least by most accounts, appears to have laid to rest the idea that our universe is either eternal or necessary. Scientific skepticism throughout the modern period had deemed it more natural to assume that the cosmos is unoriginated, but twentieth-century developments in cosmology have shaken this belief to the point where it has now become quite unfashionable to deny the Big Bang origin of the universe. Nevertheless, nature still presents itself as so vast and resourceful that it is quite tempting for naturalists—even to this day—to make the physical universe the ultimate foundation and framework of life and thought. In their view the cosmos needs no creative ground beyond itself. For many scientifically educated people today, postulating a creator distinct from nature seems superfluous. To them nature is enough, and naturalism is a reasonable worldview.

In my own conversations with scientists and philosophers I have found that naturalism is a powerfully tempting belief system. Scientific naturalists these days are usually aware of the fact that the Big Bang universe shows clear marks of having had a beginning, of being destined for eventual extinction, and hence of being only temporarily able to support life. But recently naturalists, aided by contemporary cosmological calculations, have taken to speculating that our own universe is only one episode or branch of a much larger, and endlessly inventive, set (or series) of universes.[8] Perhaps the Big Bang

---

6. See Keith Ward, "God as a Principle of Cosmological Explanation," in *Quantum Cosmology and the Laws of Nature*, ed. Robert John Russell, Nancey Murphy, and C. J. Isham (Notre Dame: Vatican Observatory and University of Notre Dame Press, 1993), 248-49.

7. Paul Tillich, *Dynamics of Faith* (New York: Harper Torchbooks, 1958), 85.

8. Martin Rees, *Our Cosmic Habitat* (Princeton: Princeton University Press, 2001); see also Lee Smolin, *The Life of the Cosmos* (New York: Oxford University Press, 1997).

universe, in all its immensity, is still only one small domain in a virtually unlimited "multiverse."

Even if the phantasm of a multiverse turns out to be correct—as well it might—it is hard not to notice the inherently religious character of such an extravagant conjecture. Like the rest of us, even scientific naturalists have a taste for eternity. If they do not expect to survive themselves, they enjoy the prospect that the larger array of supposed universes will continue cycling on indefinitely. For naturalists the power of being that Christians identify with God the Creator is translated into (or symbolized by) an endlessly resourceful cosmic matrix, a mother universe spewing out offspring universes without restraint.[9] If there is room in scientific naturalism for a sense of transcendence, perhaps it consists of stretching our minds and imaginations beyond the limits of the Big Bang universe to the mystery of an indefinitely larger set of worlds. In making this leap of faith—one that so far has no empirical basis—the naturalist displays a thirst for transcendence that rivals even the most credulous excursions in the world of religion. The fact that there is no positive evidence to support the multiverse idea makes it all the more evident that even among naturalists a religious search for infinite horizons is irrepressible.

Of course, the existence of a multiverse would be perfectly consistent with a Christian theology of creation, especially if its ultimate ground is as abundantly generous as the God of the Bible is said to be. An expansive theology of nature is open to the possibility that there exist many worlds beyond the observable one. God is the Creator of all things invisible as well as visible. And it may eventually be possible for science to predict the existence of other universes in a manner comparable to the way in which physicists, prior to actually having direct evidence of black holes, can nonetheless make a solid case for their existence. In any case, even if trillions of other worlds exist, to Christian faith there is still only one totality of created being, and this massive multiverse could never rival in breadth and depth the dimensions of infinity that we attribute to God.

The issue, then, is not whether people have a longing for the infinite. Even naturalists apparently do. The question is whether they should identify the desired infinity with the universe (or multiverse) itself, or instead with an endlessly resourceful Creativity that underlies the universe(s). Naturalists have a distaste for the latter notion, not because they lack a human thirst for the infi-

---

9. See, e.g., John Gribbin's discussion of Lee Smolin in *In the Beginning: After COBE and before the Big Bang* (New York: Little, Brown, 1993), 252-55; Andrei Linde, "Inflationary Cosmology and the Question of Teleology," in *Science and Religion in Search of Cosmic Purpose*, ed. John F. Haught (Washington, D.C.: Georgetown University Press, 2000), 1-17.

nite but because the theistic attribution of "personality" to God seems to make God smaller than the universe or multiverse.[10] Moreover, in their often pantheistic predilection for an infinite universe they sometimes interpret the idea of a beginning in time as a diminishment of the world that they had earlier assumed to be infinite in time and space.

It is hard for many Christians to understand why the belief that the natural world had a sharply defined point of origin has not been appealing to everyone. But if a person has a taste for infinite mystery, on the one hand, and understands the biblical God to be smaller than the universe, on the other, then the idea of a finite age to the universe will be taken as a religious insult. At first even Albert Einstein thought the idea of a discrete cosmic beginning would amount to a lessening of the endlessly existing universe that had become fixed in his mind and sensibilities. Until his own equations proved otherwise, he vehemently repudiated what he took to be the outrageous notion that the cosmos has not existed forever. Furthermore, it seemed to Einstein that if scientific laws are to be trusted as the basis for precise prediction, at the very least they would have to be timeless. So he questioned how science could ever be trusted if nature's unchanging habits had a merely temporal origin. If they are not eternal, the laws of science would seem arbitrary and unreliable.

Einstein himself had an unwavering conviction that the universe is a great mystery, one whose dimensions are unfathomable even by science. In cultivating a sense of mystery he was happy to admit that he was a deeply religious man.[11] The fact that the universe is intelligible at all is a great mystery that should leave scientists humble and awestruck. Without a trust that there is an uncanny coherence to the cosmos, science could never even get off the ground, but science itself can never say why the universe is intelligible in the first place. Einstein's trust in the world's eternal intelligibility, a great and incomprehensible mystery itself, made the news of a cosmic beginning repugnant to him, at least initially. For if the cosmos had a beginning, this would imply that the laws of nature are not eternally necessary, but temporally contingent instead. Therefore, science would no longer have as solid and permanent a grounding as Einstein had assumed to be the case. Consequently, after being informed by other scientists and mathematicians that his own general theory of relativ-

---

10. For example, Ursula Goodenough, *The Sacred Depths of Nature* (New York: Oxford University Press, 1998); Chet Raymo, *Skeptics and True Believers: The Exhilarating Connection between Science and Religion* (New York: Walker, 1998).

11. Albert Einstein, *Ideas and Opinions* (New York: Modern Library, 1994), 11-12.

ity implied a temporally finite cosmos, Einstein adjusted his calculations arbitrarily, hoping to preserve thereby the comfortable idea of an eternal and essentially unchanging universe. A Christian theology of nature cannot but appreciate Einstein's religious sense of infinite mystery. However, the logic of Christian faith shifts the horizon of the infinite away from an indefinitely extended cosmic sameness to the endless openness of an always renewing future.

Einstein eventually realized his mistaken calculations, but many of today's physicists still share his aversion to the idea that nature is finite in age. While some speculate that our own universe is only one in a whole batch of universes, others propose that there is no strictly specifiable temporal point of cosmic origins after all. If there is no first moment, there is no need, they claim, to appeal to a creator who started things off. If the universe never literally *began* in time, perhaps there is a sense in which it always was and always will be. And if it has no crisply definable instant of origin, what point is there in bringing up the idea of a creator? In *A Brief History of Time* the famous cosmologist Stephen Hawking states that "so long as the universe has a beginning, we could suppose it had a creator. But if the universe is completely self-contained, having no boundary or edge, it would have neither beginning nor end: it would simply be. What place then for a creator?"[12]

SCIENTIFIC BACKGROUND

The unprecedented astrophysical discoveries of the last century have by now led most scientists to doubt that our own Big Bang universe has in fact existed forever. Recent cosmology has shown that this universe's temporal duration, though unimaginably immense, is still finite. The universe had a beginning. Hence it may seem, at least at first sight, that cosmology is doing a favor for theology by positing the universe's origin in time. The Big Bang starting point understandably conjures up such questions as these: if the universe has not existed forever, does it not require a cause that exists independently of the universe? And would not this cause be equivalent to what theists call God the Creator?

Scientific suspicion that the universe had a beginning was awakened early last century especially by clear evidence that galaxies, at least on average, are

---

12. Stephen W. Hawking, *A Brief History of Time* (New York: Bantam Books, 1988), 140–141; see also Paul Davies, *The Mind of God: The Scientific Basis for a Rational World* (New York: Simon & Schuster, 1993), 66.

moving away from one another. Since they are moving apart now, one must assume that at an earlier time in cosmic history they were closer together, and a time before that when they were closer still. As scientists move back in time, retracing the lines of cosmic expansion, they conclude eventually that at some time in the very remote past, the whole of physical reality was compressed into an incalculably small, hot, and dense grain of energy/matter. It was from this primordial "egg" that the universe was hatched. As it gradually cooled down, a long sequence of events led in time to the successive emergence of free hydrogen atoms, stars, galaxies, carbon, life, and persons with minds inquisitive enough to ask where they came from and where they will end up.

The scientific consensus today is that our vast universe can be thought of as finite in age and spatial extension. As I mentioned above, Einstein himself could not initially embrace such an idea, even though his own equations should have convinced him otherwise. In 1917 the Dutch physicist Willem de Sitter, while studying Einstein's freshly drawn equations on general relativity, came to the conclusion that the new theory implied a changing, expanding cosmos rather than the eternally fixed one that Einstein had favored. A static and eternal universe would have collapsed gravitationally upon itself ages ago, but obviously it has not done so. That the universe is in fact continuing to expand spatially could only mean that there had to be a beginning at some time in the far distant past. Since gravity tugs on everything, if the universe had existed forever every material body in it would have bunched up with all the others by now. But since this has not happened scientists should already have suspected even prior to Einstein that the universe is not infinitely old.

At any rate, in 1922, the Russian mathematician Alexander Friedmann concluded that Einstein's own theory of general relativity is incompatible with the idea of an eternal, essentially unchanging universe. Around the same time, the Belgian priest and physicist George LeMaitre proposed to Einstein that the universe of general relativity must have sprung from an infinitesimal physical speck in an originating event that the cosmologist Fred Hoyle would later sarcastically refer to as the "Big Bang," a label that persists. Having learned of de Sitter's, Friedmann's, and LeMaitre's interpretations, Einstein first rejected their conclusions; he introduced a hypothetically repulsive force into his theory that would counter the pull of gravity and explain why the heavenly bodies have always remained so spread out.

Einstein can be pardoned for this error of judgment, since early on he was unaware of the amazing new findings of the American astronomer Edwin Hubble. In one of the most important scientific discoveries of the last century Hubble and his assistants were busy gathering observational evidence that the universe is changing rather than staying stuck in perpetual immobility. Not long after the discovery of galaxies early in the twentieth century evidence arrived that many of these "island universes" were moving apart from one another. The "red shift" in spectroscopic analysis of light waves emanating from many galaxies meant that they are moving away from each other. Their accelerating rate of speed was later dubbed the "Hubble Constant," and it is one of the most important, though still disputed, numbers in contemporary cosmology.

Thoughtful scientists could in principle already have predicted in yet another way that the universe is finite in age. For if the supposed billions of stars in the sky had been fixed in place for all eternity rather than moving away from each other, light from all of them would have reached our planet by now. This bombardment of photons would be flooding every earthly night with dazzling illumination. Darkness would not exist here. And yet it does. This puzzle, known as Olber's paradox, was not solved until the time of Einstein and Hubble, when science could show that the universe is finite in age and that there has not yet been enough time for stellar light to drown out the night. Until recently, astronomers had not calculated that nocturnal darkness here depends on the fact that the universe is finite in temporal duration and that light-emitting bodies are separating from one another at an accelerating rate of speed.

At present the expanding universe predicted by Einstein's equations appears to have received empirical confirmation. Nevertheless, even after Einstein's theory and Hubble's findings, misgivings about the Big Bang persisted. Some scientists found it more convenient to stick with the idea of an eternal and stable universe. Although in 1965 the Bell Laboratory scientists Robert Wilson and Arno Penzias had discovered in the cosmic background microwave radiation what turned out to be smoking gun evidence of the Big Bang, doubts lingered on. A major reason for this skepticism was that scientists assumed that the Big Bang universe should have expanded more uniformly than it has, whereas astronomy shows that the visible matter in the universe is distributed unevenly. Galaxies, stars, planets, gases and other kinds of matter are scattered in such a way that immense empty spaces lie between some clusters while smaller distances divide others. How could this be so unless there had also been irregularities in the infant universe?

To account for the absence of uniformity that astronomy actually records in the present universe, Big Bang proponents would have to demonstrate that the newborn cosmos already contained in seedling form the unevenness evident today. But how could the state of the universe that long ago be measured some fourteen billion years later? It is the mark of science's power of discovery and ingenuity that it can now answer this question. In the spring of 1992 it was announced that careful measurement of the microwave temperature differentials in outer space by the COBE (Cosmic Background Explorer) satellite provides powerful evidence for the Big Bang. The satellite had detected slight differences in temperature in the still remaining, though now quite cold, after-glow of the radiation that accompanied the Big Bang. The roots of this unevenness extend back in time to the earliest stages of the universe's existence. The Big Bang theory now seems quite safe.[13]

THEOLOGICAL IMPLICATIONS?

These days scientific consensus overwhelmingly supports Big Bang cosmology. But does scientific evidence for the Big Bang render the theology of creation any more intellectually respectable than before? Has modern science finally reached conclusions about nature that make a creation theology more credible perhaps than ever?

The history of theology's engagement with science advises caution in responding to such a question. At this point, therefore, it is appropriate to provide a brief typology of five different ways in which various parties may approach not only the question of creation and the Big Bang but also other questions in science and theology.[14]

1. Conflation. The first approach is one that fails to distinguish science and theology clearly. It fuses or "conflates" one into the other. Conflation is the careless overlapping of ways of understanding that are logically distinct and that rely on methods of understanding that have different objectives and use different kinds of evidence. As I noted earlier, for example, science tries to solve problems, whereas theology guides us into mystery. But conflation overlooks such distinctions either by forcing science to answer religious questions, or expecting sacred texts or theology to provide scientific enlightenment. A good example of conflation is a literalist reading that takes the book of *Genesis* to be an authoritative source of *scientific* information rather than an invitation to a radically new understanding of God, being, and ourselves. Many biblical liter-

---

13. A good introduction to the story of the scientific discovery of the Big Bang, prior to the COBE project, is Timothy Ferris, *Coming of Age in the Milky Way* (New York: Harper Perennial, 1988).

14. In my book *Science and Religion: From Conflict to Conversation* (Mahwah, N.J., and New York: Paulist, 1995), I developed a typology involving four distinct ways of relating science to religion. Here I am slightly modifying my earlier presentation by making "conflation" a part of the typology rather than an approach that falls outside of it.

alists, especially those known today as "creation scientists," read the Bible as though one of its functions is to provide a *scientific* account of origins.

Conflation is the consequence of a failure to distinguish scientific method carefully from a worldview or belief system. Not surprisingly, it is the root cause of many difficulties in the historical relationship of science to theology. However, it is not only religious fundamentalists who have indulged the conflationist impulse. Many of those I have been calling scientific naturalists also tend at least implicitly to commingle science with their (unverifiable) belief that the universe can be understood completely in a materialist way. For example, evolutionists such as Stephen Jay Gould, E. O. Wilson, and Richard Dawkins superimpose their belief that "matter is really all there is" directly onto evolutionary biology, scarcely aware that in doing so they are leaving science behind and entering the world of ideology.[15] The materialist spin they put on science is no less conflationist than is creationism or what has come to be known as "intelligent design theory."

2. Conflict. A second way of relating science to belief is to think of them as irreconcilably opposed. This is the *conflict* position, the one held by most scientific skeptics today. Religious believers, for their part, endorse the conflict position whenever they take passages from Scripture or specific teachings of their faith and place them in opposition to science. However, in this book when I speak of the conflict position I am thinking primarily of scientific naturalism, the belief system that rejects the claims of theology because the latter cannot be tested by science.

The conflict position holds that science is the only trustworthy way to understand anything at all. So, since belief in God or Jesus' resurrection is scientifically untestable, the conflict approach insists that intellectually honest people should refrain from embracing these fundamentals of Christian faith. Only those claims that can pass muster as scientifically accurate are acceptable. All the fuzzy teachings of Christianity and other religions only lead the honest seeker of truth away from the real world. Christian faith at best can have only a moral or emotive significance. It can have no cognitive standing.[16]

Those who view theology as an enemy of science often think that biblical revelation should be rejected because it does not dish out reliable scientific

---

15. Stephen Jay Gould, *Ever Since Darwin: Reflections in Natural History* (New York: W. W. Norton, 1977), 12–13; Richard Dawkins, *River Out of Eden: A Darwinian View of Life* (New York: Basic Books, 1995); E. O. Wilson, *Consilience: The Unity of Knowledge* (New York: Knopf, 1998), 6.

16. I shall return to the question of revelation and the criteria of truth in chapter 10.

information. For example, the evolutionists E. O. Wilson and Daniel Dennett both dismiss theology for taking seriously ancient Scriptures that have nothing to say about evolution.[17] Both Dennett and Wilson think of themselves as sophisticated thinkers, but their criticism exhibits exactly the same biblically literalist assumptions as those of their creationist opponents. That is, they take it for granted that if the Bible is an authoritative source of truth it should be *scientifically* accurate. The shallow literalism that leads creationists to reject evolution as unbiblical, for example, also ironically moves Daniel Dennett to declare that "Darwin's idea has banished the book of Genesis to the limbo of quaint mythology."[18] This highly respected philosopher assumes, along with religious fundamentalists, that the Bible should have been dishing out reliable scientific information all along. Meanwhile, Wilson insists in the same incongruous way that if the Bible were truly revelatory it should have gotten its science straight. And since neither Dennett nor Wilson can find in Genesis any helpful insights about evolution, they declare the Scriptures untrustworthy.

Once, on a panel made up of theologians and scientific skeptics, I was scolded by a fellow participant for taking Genesis seriously since it is so full of "lies" as judged from a scientific vantage point. Creationists, in other words, do not have a monopoly on literalism. Catholics can cite the fact that even as conservative a document as Pope Leo XIII's encyclical *Providentissimus Deus* (1893) gave clear instructions to the faithful not to look for scientific truths in the Bible.[19] But apparently no parallel message has penetrated the world of materialist evolutionists. In alloying science, and especially evolutionary biology, with scientism and materialist naturalism they present to their readers an amalgam of belief and science that rivals other expressions of fundamentalism in its failure to distinguish science from belief systems. The conflict approach is still mired in the conflationist swamp.

3. Contrast. Conflict, as we have just seen, is just the other face of the conflationist coin. A third and much more elegant approach avoids conflict by first avoiding conflation. I call this third approach *contrast* since it takes pains to

---

17. Wilson, *Consilience*, 6; and Daniel Dennett, as interviewed in John Brockman, *The Third Culture* (New York: Touchstone, 1995), 187.

18. Dennett, as interviewed in Brockman, *Third Culture*, 187.

19. "It should be remembered that the sacred writers, or more truly 'the Spirit of God who spoke through them, did not wish to teach men such truths (as the inner structure of visible objects) which do not help anyone to salvation'; and that, for this reason, rather than trying to provide a scientific exposition of nature, they sometimes describe and treat these matters either in a somewhat figurative language or as the common manner of speech those times required, and indeed still requires nowadays in everyday life, even amongst most learned people" (Pope Leo XIII, *Providentissimus Deus* 18).

distinguish science sharply from any worldview whatsoever, whether theological or naturalistic. Repudiating both the conflationist and the conflict approaches, the contrast approach is attractive to many thoughtful and logically gifted scientists and theologians who rightly understand the need to make important distinctions between scientific and theological inquiry. Contrast argues that theology and science are addressing radically disparate sets of questions, and so any real conflict between them is impossible. Theology and science both lead to truth, but by different roads. Because they are so different in their methods, and have such distinct objectives, it makes no sense to place one in competition with, or opposition to, the other.

The contrast approach typically holds that science asks *how* things happen in nature, whereas theology is concerned with such questions as *why* there is anything at all rather than nothing. Science is about natural, physical causes; theology is in search of *meaning*. Science deals with solvable *problems;* theology with limit questions that introduce us to the *mystery* of God. Science answers specific questions about the *workings* of nature, whereas theology expresses concern about the ultimate *ground* of nature. Only by sequestering science and theology in separate camps can skirmishes between them ever be prevented or resolved.[20] The whole ugly affair involving Galileo and the church, for example, could have been avoided if theologians and philosophers had not encroached upon the newly hatched autonomy of empirical science.

4. Contact. The contrast approach is uncompromising in its rejection of conflation. It rightly acknowledges that the conflict approach would never have any appeal if science had not been mixed up in the first place with worldviews, whether religious or naturalistic. Contrast is like a cold bath for those who have caught the conflationist fever. But the clinically neat contrast position is too simplistic to match the world of actual human thought and inquiry. One must not conflate science with belief, of course, but science nevertheless always has implications for the world of religious belief and theology. I shall be advocating, therefore, what may be called the *contact* approach as the one that a theology of nature must follow most closely. Contact forbids any confusion of science with religion, but it also recognizes that it is impossible to isolate theology absolutely from the results of scientific discovery. It is realistically aware of the conflationist temptation, but also of the fact that various belief systems cannot help being affected in some way or other by science. Contact accepts the crisp distinctions that the contrast position makes, but it

---

20. A good example of a thoroughgoing "contrast" approach is Stephen Goldberg, *Seduced by Science: How American Religion Has Lost Its Way* (New York: New York University Press, 1999).

is convinced that we must distinguish science and religion only in order to *relate* them more meaningfully to each other.

The task of the contact approach is a delicate one and must be done over and over in each new epoch of scientific discovery. But it is unrealistic to deny that one's general understanding of what is true and important theologically will be affected by what is generally assumed to be true and important scientifically. At times the contact of faith and theology with science will be painfully confusing, as in the church's encounters with Galileo's cosmology or Darwin's picture of life. Nevertheless, Christian theology, in order to demonstrate its own commitment to right understanding, cannot ignore new developments in the natural sciences. After theologians have listened to Darwin, Einstein, and Hubble, they cannot have exactly the same thoughts about creation or the Creator that they had before. One of the main tasks of a theology of nature, therefore, is to inquire into the possible religious meaning for Christian faith of natural events and scientific ideas such as the Big Bang.

5. Confirmation. Finally, a fifth approach emphasizes the ways in which Christian beliefs have prepared the soil for science's flourishing, both historically and logically. For lack of a better mnemonic I refer to this approach as *confirmation*. One meaning of the verb "to confirm" is that of providing support or strength. What I propose then is that although Christian theology cannot and should not attempt to answer scientific questions, nevertheless it can quietly *confirm* the whole scientific enterprise, lending its support to open-ended research. For example, Christian teaching gives a very good answer to the limit question: why should one bother to do science at all? The answer is that "truth is always worth seeking," and in support of that claim one may cite the Christian conviction that the universe is grounded in the font of wisdom, meaning, and truth that we refer to as God. When the divine Word, on which the whole of creation is modeled, becomes incarnate, the entire world is revealed to be inseparable from an eternal principle of intelligibility. Theologically speaking, science is justifiable only if the world it explores is intelligible. Only the scientist's spontaneous trust that the universe is intelligible can launch a life of research. But science itself cannot justify that trust. It is the task of theology to do so, and if it is successful it has the effect of supporting, or confirming, the scientific journey of discovery.

CONFLATION AND CREATION

Conflation, conflict, contrast, contact, and confirmation are five distinct ways of relating science to theology. Let us now take a brief look, therefore, at five corresponding ways of thinking about the implications of current cosmology for a Christian theology of creation.

To start with, how would the *conflationist* approach react to news about the Big Bang? Typically it tends, in my view, to be overly excited. Norman Geisler and Kerby Anderson, for example, are so ebullient as to announce that the Big Bang theory "has resurrected the possibility of a creationist view of origins in astronomy."[21] A somewhat more subdued, but still overblown, proposal comes from the astronomer Robert Jastrow. Even though he had earlier professed to being an agnostic, Jastrow nevertheless implies in his popular book *God and the Astronomers* that the cosmological synthesis provided by Einstein, Hubble, and LeMaitre lends surprising support to the doctrine of creation. In a spirit of resignation Jastrow confesses that the Big Bang theory comes as a complete shock to naturalistic suppositions. Now, after several centuries of embarrassment at the hands of science, theologians at last may take heart:

> At this moment [as a result of Big Bang cosmology] it seems as though science will never be able to raise the curtain on the mystery of creation. For the scientist who has lived by his faith in the power of reason, the story ends like a bad dream. He has scaled the mountains of ignorance, he is about to conquer the highest peak; as he pulls himself over the final rock, he is greeted by a band of theologians who have been sitting there for centuries.[22]

A theology of nature, however, would advise readers to be skeptical of such conflationist overtures. Drawing theological conclusions directly out of science has a long but highly suspect history. Isaac Newton, for example, was convinced that his physics had finally given theology a solid intellectual foundation, but the marriage of science and theology that he envisaged has not been a successful one. As soon as physics had established itself as an autonomous discipline, the earlier appeals by natural philosophers to the idea of God in order to prop up the Newtonian world system began to seem superfluous.[23] In a parallel way in 1951, Pope Pius XII tried to convince a gathering of scientists that the fledgling Big Bang theory adds plausibility to the doctrine of creation.[24] More recently a number of believing scientists have

---

21. Norman L. Geisler and J. Kerby Anderson, "Origin Science," in *Religion and the Natural Sciences*, ed. James E. Huchingson (New York: Harcourt, Brace, Jovanovich, 1993), 202.

22. Robert Jastrow, *God and the Astronomers* (New York: W. W. Norton, 1978), 116.

23. See Michael J. Buckley, S.J., *At the Origins of Modern Atheism* (New Haven: Yale University Press, 1987).

24. See Ernan McMullin, "How Should Cosmology Relate to Theology?" in *The Sciences and Theology in the Twentieth Century*, ed. A. R. Peacocke (Notre Dame: University of Notre Dame Press, 1981), 17-57.

argued that astrophysics provides a solid reason to accept God's existence. But one must be on the lookout for the implicit overlapping, and hence confusing, of scientific and religious modes of understanding in such proposals. It is safer to shiver in the icy waters of contrast than burn in the fires of conflict and conflation. As Paul Tillich advises, theologians should never embrace any scientific ideas for purely theological reasons.[25]

What, then, are the alternatives to the conflationist synthesis of Big Bang cosmology with creation theology? Each of the other four interpretations—those I am calling conflict, contrast, contact, and confirmation—has its own response.

CONFLICT

Naturalism claims that science is incompatible with theology, or at least that scientific discoveries increasingly render theological claims obsolete. On the topic of cosmos and creation, it argues that even if the universe had a beginning in time, this does not require that it have a cause outside itself. Quantum physics, it is claimed, allows for the possibility that the universe came into being *without any cause*. The cosmos possibly had a beginning, but recent physics says that it could have popped into existence spontaneously, that is, without any antecedent cause. Hence there is no need for a creator.

This is the sort of claim that a good many religiously skeptical cosmologists are making today. The scientific basis for their thinking is roughly as follows. At one time, according to recent cosmological speculation, the universe was no larger than a subatomic particle, and at that instant it must have behaved the way all such infinitesimal particles do. Quantum mechanics allows that the emergence of such particles does not need to have any determining cause whatsoever. The so-called virtual particles of microphysics simply pass in and out of existence spontaneously, without any specifiable cause. The ordinary laws of physical causation do not apply in the region of the very small. Consequently, given its diminutive dimensions, the primordial cosmic particle could have popped into existence spontaneously. As Douglas Lackey says, "the big bang might have no cause," having come into existence "from a vacuum, that is, *from nothing*."[26]

---

25. Tillich, *Dynamics of Faith*, 85.

26. Douglas Lackey, "The Big Bang and the Cosmological Argument," in *Religion and the Natural Sciences*, ed. James E. Huchingson (New York: Harcourt, Brace, Jovanovich, 1993), 194 (emphasis added).

As I noted above, the renowned astrophysicist Stephen Hawking provides another interpretation, but it is one that also professes to allow for a purely naturalistic understanding of Big Bang cosmology. Although we now have to admit that the universe is not infinitely old, Hawking proposes, it does not necessarily have a sharply defined temporal beginning either. The eminent cosmologist speculates that time has emerged only gradually, rather than suddenly, out of a primordial space-time matrix devoid of duration. The universe may never have leapt abruptly into existence at any definable first moment since there *was* no first moment. And if there was no first moment, there is no need to look for a first cause.[27] So Big Bang cosmology provides no obvious support for Christian faith. My own response to the conflict position unfolds in the following discussion of contrast, contact, and confirmation.

CONTRAST

The contrast approach insists that Big Bang cosmology proves nothing one way or the other as far as creation is concerned. Contrast insists that those who embrace the doctrine of creation should be loath to look to physics for support. The temptation to read theological implications directly out of cosmological discoveries is a constant one, but for the sake of theology's integrity one should resist it uncompromisingly. Tying the doctrine of creation as closely to Big Bang theory as Jastrow and Pope Pius XII want to do would leave theology in an awkward spot if it ever turned out that Big Bang theory is off the mark, or if the idea of a multiverse or mother universe could be verified at some remote time in the scientific future. The contraster's point is that theology should never appeal for its credibility to any particular scientific discovery. After all, science is habitually changing its mind, and the theology that binds itself in marriage to the science of today can easily become widowed tomorrow. I believe that a contemporary theology of nature must take such a prospect into account, even as it also seeks a closer conversation with science.

The contrast strategy is to distinguish theology, both in method and content, from each and every scientific idea. This severe sequestration precludes any opportunity for the two to tangle. Even though Big Bang theory now carries the day in science, contrasters caution that theologians should not derive any comfort from it. Moreover, it seems somewhat impious to base the plausibility of the venerable doctrine of creation on the variable versions of con-

---

27. Hawking, *Brief History of Time*, 140-41.

temporary physics. According to the strict rules of the contrast approach, theology should not fall for the seductions of Big Bang cosmology even if there are prima facie affinities between Genesis and the new astrophysical portraits of the world's beginning. Biblical religion and astrophysics, after all, are talking about two entirely different sets of truths, and the plausibility of the doctrine of creation depends in no way on the veracity of Big Bang theory. Were we to read in tomorrow's newspaper that the theory is flawed or scientifically mistaken, it should make no difference to a well-grounded theology. Big Bang physics has nothing to say about the true meaning of creation, nor do religious accounts of creation provide any useful information about the physical origins of the universe.

Big Bang cosmology is quite possibly a rock-solid scientific theory, but according to the contrast position, the doctrine of creation is about much more than cosmic beginnings. To begin with, it is a response to the question: why is there anything at all rather than nothing? Creation is not so much about chronological origins as about the world's ontological dependence on a beneficent principle of being that exists independently of the cosmos. Creation theology's concern is that of awakening people to the divine graciousness that allows something other than itself to exist at all. In St. Bonaventure's terms, a theology of creation leads the mind to an awareness and gratitude for the "fountain fullness" (*fontalis plenitudo*) and overflowing goodness from which the world's being arises.[28]

The doctrine of creation should lead us to acknowledge, and be grateful to, the power of being that grounds, sustains, and renews the world. In order to experience the power of divine creativity here and now it is especially important to avoid identifying creation exclusively with an originating moment in the distant past. As the theologian Keith Ward rightly emphasizes,

> it is wholly inadequate to think of God having created the universe at some remote point of time—say, at the Big Bang—so that now the universe goes on existing by its own power. This popular misconception, that "the creation" is the first moment of the space-time universe, and that the universe continues by its own inherent power, wholly misconstrues every classical theistic tradition. It is irrelevant to a doctrine of creation *ex nihilo* whether the universe began or not; that the universe began was usually accepted because of a particular reading of Genesis 1. The doctrine of creation *ex nihilo* simply maintains that there is nothing other than God from which

---

28. Bonaventure's expression "fountain fullness" is examined carefully in Ilia Delio, OSF, *Simply Bonaventure: An Introduction to His Life, Thought, and Writings* (New York: New City Press, 2001).

the universe is made, and that the universe is other than God and wholly dependent upon God for its existence.[29]

Thus, Stephen Hawking is way off the mark, although he is not alone, in associating creation with beginnings. This confusion is what allows him to eliminate a creator, for by smearing out any sharp point of cosmic beginnings, he supposes that there is nothing left for a creator to do except start things off. Yet even an eternally enduring universe would require, in order to sustain and renew its existence, a primordial and persistent "fountain fullness," an endlessly deep wellspring of being. It is in an ontological rather than exclusively chronological way that one must understand God's creativity. Moreover, locating the act of divine creation exclusively at the temporal onset of the universe can lead to deism, an emaciated version of theism according to which God is relegated to being only the first cause in a series of natural events and who becomes increasing irrelevant to the world's existence here and now.

It follows, therefore, that theology does not need to worry about what appears to science to be the spontaneous origin of the universe as this is depicted by those familiar with quantum physics. The suggestion by some scientific naturalists that the infinitesimal early universe just popped into existence, suddenly and uninvited, out of a vacuum matrix, scarcely accounts for why there is being rather than nothing. After all, even the quantum vacuum, the field of potentiality out of which the universe is said to have arisen, is not the same thing as naked nothingness in the sense of the *ex nihilo* referred to by creation theology. The simple fact that matter-energy exists, no matter how wispy it seems to common sense in some of its more subtle states, is still enough to evoke the question of why those states of being exist at all. Tracing events back either to a first efficient cause, to a vacuum matrix, or to a spontaneously broken perfect symmetry, or even further back into the mists of an eternal past, does not preclude an ultimate power of being that continually and graciously sustains the world's existence as a whole while also opening it up to a new future.

CONTACT

I have just summarized the contrast approach to a theology of creation. Many Protestant and Catholic theologians are attracted to it because of the apparently clean way in which it detaches Christian faith from what happens in the

---

29. Keith Ward, "God as a Principle of Cosmological Explanation," 248-49.

world of science. As long as science and theology are seen as responding to diverse sets of questions, there can be no conflict between them. But should theology stop here? Perhaps the contrast approach is too sanitary. Even though the crisp segregation of science from theology is logically compelling and avoids the nuisance of conflation, I am convinced that a more adventurous interaction is necessary. Indeed, a theology of creation is needlessly impoverished if it remains indifferent to the fascinating developments in cosmology over the last century. Theology need not play it so safe that it forfeits the opportunity to grow.

The *contact* approach that I will sketch in this section seeks to avoid any new conflation, and it agrees that science and theology are logically distinguishable. But it also allows cosmology to have an impact on theology. In fact, a Christian theology of creation may receive new life and meaning by engaging in conversation with current cosmology. Theology need not base itself directly on physics, but shielding itself from cosmology altogether only lends support to the privatizing and hence diminishment of faith. Big Bang science can enlarge our sense of creation and in doing so expand our sense of God. How so?

Even though the Nicene Creed refers to God the Father as Creator of the heavens and the earth, theology has remained so tied up with questions about human existence, its history, and its hurts, that in modern times it has generally failed to notice how intricately our being is linked to the natural world with its own fascinating history and indeterminate future. Partly due to the influence of the philosopher Immanuel Kant (1724-1804) the universe came to be thought of as a kind of construct of the human mind, or as a mere backdrop for the human drama, rather than as a set of objects deserving of formal study itself. Accordingly, the universe became virtually lost to modern theology, which until recently had itself become increasingly anthropocentric.

The new cosmology, however, is theologically significant by virtue of the fact that it has brought the whole universe into the foreground again. The physicist and Catholic scholar Stanley Jaki rightly points out that current cosmology, starting especially with Einstein, has "restored to the universe [the] intellectual respectability which Kant had denied to it."[30] Einstein's new interpretation of gravity proposed that the cosmos is a finite set of interrelated things, and that it calls for a focal attention that Kant could not give it. While pushing the universe into the background, Kant's obsession with human subjectivity has had the effect of decosmologizing modern thought, including theology.

---

30. Stanley L. Jaki, *Universe and Creed* (Milwaukee: Marquette University Press, 1992), 27.

However, after Darwin, Einstein, and quantum physics the human subject can no longer be so easily separated from the universe. It turns out that the universe is much more than a timeless setting for the human adventure. In fact, the universe is itself the principal creative adventure, and there is no reason to assume that our species is the sole reason for its existence. Even on earth we are not the only species of life, and for all we know the two or three hundred billion galaxies in the observable universe may harbor numerous small oases of intelligent life. In any case, we cannot go wrong either religiously or scientifically if our first response to the fact of our existence is one of gratitude that we have been invited to be at least a small part of an immense *cosmic* journey. We are a very important part of the universe, but we are not the whole story. Increasingly the questions of who we are, where we came from, what we must do, and what we may hope for can be answered only if theology attends to what is going on in the universe at large. And it cannot do this without contact with science.

Big Bang cosmology implies that the universe, immense and old though it may be, is nevertheless spatially and temporally finite. But to acknowledge that the universe is finite is also to realize that it is *contingent*. In other words, there is no a priori necessity for its existing at all, nor for its being precisely the kind of universe it is. So *why* does it exist if it does not have to? And why is it the kind of universe it is, when other kinds of universes are physically conceivable? Such questions arose long before Einstein, of course, but in Christian theology after Kant it has been the existence and meaning of the human subject rather than the entire universe that has held center stage. To a great extent this is still the case, but current cosmology has begun to shift the questions from why I am here to why the universe is here and why it is able to give rise to life, mind, and persons.

In contemporary scientific thought one may now and then pick up strains of a fresh sense of awe that the universe exists at all and that it is the kind of universe that can produce living and thinking beings. Stephen Hawking, for example, appears to be calling on a metaphysical principle that can "breathe fire" into the physicist's equations so as to allow for the actual existence of a cosmos. Why, he asks, does the universe go to all the bother of existing at all?[31] Since creation faith arises from wonder at the mystery of being, it is of interest to theology that even scientists such as Hawking are experiencing what Paul Tillich calls "ontological shock."[32] That is, they are struck by the fact *that*

---

31. Hawking, *Brief History of Time*, 174.
32. Tillich, *Systematic Theology*, 1:113.

anything exists at all. This kind of wonder provides a good reason for theology to appreciate the discoveries of science.

Furthermore, the scientist's curiosity about cosmic origins is inseparable from the nearly universal religious interest in origins. Even though the theologian accepts the contraster's distinction between creation theology and Big Bang cosmology, it is artificial to separate altogether a scientist's personal enthusiasm about discovering cosmic origins from a more general human and religious concern for beginnings. Granted that scientific cosmology is logically distinguishable from religious cosmogony, nonetheless the two kinds of inquiry have a common root in our personal preoccupation with the perennially haunting question of where everything comes from ultimately.

Finally, and most significant for the theological themes I am highlighting in this book, I would emphasize that the new scientific cosmology presents us with a universe that is still very much in the making. Paradoxically, the more science has turned its attention to a study of the remote cosmic past, the more it has opened up reasons for preoccupation with the future of creation up ahead. Together with evolutionary biology, earth science, and new developments in other areas of science, current cosmology is reinforcing the impression that the creation of the cosmos is still far from being over. In the light of science, theology may confidently affirm that *creation is still happening*. And if it is still happening one cannot but wonder where it may all be going. The more science peers into the cosmic past, as Teilhard de Chardin often observed, the more the question of the cosmic future widens out ahead of us.[33]

That we are now living in an unfinished world may not immediately strike us as terribly momentous, but in fact it means that the universe may still be very early in its full unfolding. This prospect, in turn, allows the future to take on a new significance, one that it could not have if we still impaled the universe on an older physics with its erroneous assumption of the universe's eternal sameness. Whereas science now turns our attention back toward the past for explanation in terms of what is earlier and simpler, the unfinished world-in-process that cosmology after Einstein and Hubble has dug up moves us also to look toward the *future* in order to make complete sense of it.

It is especially here that theology can meet up with cosmology today. The God of revelation is identifiable as the promise-maker who opens up a new future for the world, and as a humble, self-emptying love whose gracious self-withdrawal allows creation to come into being as something distinct from its

---

33. Teilhard wrote to a friend during one his journeys in 1923: "I am a pilgrim of the future on my way back from a journey made entirely in the past" (Pierre Teilhard de Chardin, *Letters from a Traveler* [New York: Harper & Row, 1962], 101).

Maker. The revelatory image of God's promise and descent now allows a theology of creation to make religious sense of science's relatively recent discovery that the universe is a still-unfinished story.

It is God's desire for what is truly other, a longing at the heart of the Trinity, that accounts ultimately for the fact that the universe is not rounded off in final perfection in an originating act of creation. A Christian theology of creation can confidently conjecture that God's humble and selfless love seeks something *other* than itself, an other without which God's love cannot be actualized. The created universe is grounded in the selfless love of God, and such is the nature of love that if God truly loves the world (John 3:16) then God must will the independence of creation. Thus, it should not be too surprising that the universe would be granted an immensity of space and time in which to become distinctively itself—as something other than God.

Furthermore, a theology of creation envisages the divine promise as the ultimate explanation of the anticipatory character of nature, a trait especially visible in the fact of "emergence." By emergence scientists mean that in the course of time the cosmos has occasionally produced dramatic innovations involving astounding increases in complexity and new organizational principles that were not previously operative. The most obvious instances of emergence are the birth of life and, later on, the sudden eruption of intelligence on earth, but these are preceded by less ostentatious chapters in the evolution of the universe.[34] The play of large numbers in the context of deep cosmic time and ordinary physical and chemical processes can give rise at critical points to new ways of being and functioning.[35] The emergence of such novelty is especially puzzling to science, which is constrained to explain what is new and complex only in terms of what is older and simpler. But emergent phenomena catch the eye of a scientist at all only because they are systems in which "more comes out than was put in."[36] And even though naturalists are satisfied to explain the more in terms of the less, and the later in terms of the earlier, there is an air of magic in such "explanation" if it is taken as adequate. The question then is where the new organizational principles come from.

What emergence shows, at the very least, is that what the universe is really all about cannot be determined simply by looking toward its past, as science

---

34. Harold J. Morowitz, *The Emergence of Everything: How the World Became Complex* (New York: Oxford University Press, 2002); see also Philip Clayton, *Mind and Emergence: From Quantum to Consciousness* (New York: Oxford University Press, 2004).

35. See Stephen Johnson, *Emergence: The Connected Lives of Ants, Brains, Cities, and Software* (New York: Touchstone, 2001).

36. John Holland, *Emergence: From Chaos to Order* (New York: Perseus Books, 1999), 15, 112, 225.

does, but only by simultaneously turning our eyes toward its future. "The grandeur of the river is revealed not at its source but at its estuary," Teilhard rightly insists.[37] "In nature," as Ralph Waldo Emerson adds, "every moment is new; the past is always swallowed and forgotten; the coming only is sacred."[38]

Whenever we try to grasp the true dimensions of something that is still coming to birth—such as life and thought—simple description of its past is not enough. The cosmos dissolves more and more into the incoherence of multiple, disconnected fragments if we look only in the direction of its remotest past. Intelligibility can be found only as we look toward the future integration of the world's initial multiplicity into ever-deeper unity and coherence. *To create is to unite.* Science's method of peeling back the layers of the past needs to be complemented by another kind of exploration, one whose concern is to open us to the future. If we wish to understand the universe here and now we need to look toward what it *promises* to become up ahead.

Gaining such a perspective, it seems to me, is one of the main goals of a genuinely biblical theology of nature. It is especially in our common quest for the future that we may locate much of the theological importance of conversation between theology and cosmology today. Science's habitual obsession with forcing the universe to fit only what has been learned from tunneling back toward the past has to be supplemented by theology's attention to a future that will be full of surprising twists and turns. This is why theology cannot ignore science's generous picture of a universe still surging toward ever more emergent novelty and diversity.

The idea that our universe has sprung up out of an unrepeatable singularity gives an irreversibility to time that makes every moment new, every experience fresh, and every recurrence unique. Big Bang cosmology helps theology realize that the universe is not a set of static and eternal laws but a momentous, still-unfolding story in which no chapter or no page is ever duplicated. Creation is perpetually new every day. As Teilhard says:

> The fact is that creation has never stopped. [God's] creative act is one huge continual gesture, drawn out over the totality of time. [Creation] is still going on; and incessantly even if imperceptibly, the world is constantly emerging a little farther above nothingness.[39]

---

37. Pierre Teilhard de Chardin, *Hymn of the Universe*, trans. Gerald Vann, O.P. (New York: Harper Colophon Books, 1969), 77.

38. Ralph Waldo Emerson, "Circles," in *Emerson's Essays* (New York: Harper Perennial Books, 1981), 226.

39. Teilhard de Chardin, *The Prayer of the Universe: Selected from Writings in Time of War*, trans. René Hague (New York: Harper & Row, 1968), 120-21.

My point here is that nothing could be more deadening to the human spirit, or for that matter to scientific exploration, than the assumption that everything of consequence in the story of this universe has already taken place. Unfortunately, several centuries of mechanistic science have led many scientific thinkers to suspect that the only realistic intellectual stance is a cosmic pessimism that views everything as utterly bound to the determinism of the past and to the simple elements to which science strives to reduce complex and novel phenomena. Such an outlook still thrives in intellectual circles, but the emergent features of the cosmos now render such a naturalistic bias otiose, and this cannot help but be good news to a theology shaped by a sense of reality's promise. Partly as a result of its earlier encounters with evolutionary science, but now because of its engagement with cosmology and the new scientific study of emergence, theology today has a surer sense than ever before that the cosmos is still being called into being.

Of course, a major question arises from the theological overtures of the contact approach: If the cosmos is ultimately destined to die a heat death, why should we place any hope in its future destiny? Isn't the cosmic pessimism of the pure scientific naturalist the most realistic way of understanding nature after all? I shall address this important question in chapter 9.

CONFIRMATION

A theology of creation is not only consistent with, but also supportive of, science. The doctrine of creation may even have had something to do with the development of science in the Western world, although there is much controversy about this proposal. It seems to me that, at the very least, the empirical imperative underlying science can find powerful support in the assumption that the world is a contingent creation of God. According to some historians, the biblical belief that the world was created, and hence is neither eternal nor necessary, has given a significance to empirical inquiry that alternative worldviews have not.[40] But whether this idea is historically acceptable or not, the doctrine of creation provides at least a *theological* justification of the empirical turn in modern science.

To understand this point, suppose that the natural world simply *has* to have existed from all eternity and that it *has* to be just the kind of universe it is. If

---

40. Michael Foster, "The Christian Doctrine of Creation and the Rise of Modern Science," *Mind* 43 (1934): 446-68.

the universe were necessary rather than contingent, we might wonder whether empirical investigation is worthwhile at all. Since a necessary universe would have to be the way it is, we would gain little from actually looking at it.[41] Once we have figured out the eternal laws by which it runs, we could, at least in principle, find out more about it simply by deducing its properties from first principles rather than by actually looking at it. We could do this simply with pencil and paper, or today with a computer. Empirical investigation or field-work would be unnecessary. Once we realized that every aspect of the cosmos is determined to be just what it is, actual observation of the world could be supplanted by sheer mathematical prediction, a prospect that even today some philosophers and scientists still idealize. There would be no need for a hands-on scientific examination of the world's particulars if everything that happened in it were the inevitable outcome of a series of determining causes. The empirical side of science would become increasingly superfluous the more its actual contingency dissolved into the fiction of hard necessity.

A theology of creation, on the other hand, entails the belief that the physical universe is contingent. There is no eternal necessity that it exist at all or that it be exactly the kind of universe it is. Both its existence and its specific characteristics are products of the free decision of the Creator. Consequently, we can come to know about the universe not through pure deduction but only by a combination of observation and deduction. A theology of creation, therefore, endorses (confirms) the impulse toward discovery that underlies good science. It denies that there is a rigid necessity in the existence and specific features of the universe. The universe does not *have* to be at all, nor does it *have* to be exactly the kind of universe it is. Thus, science is open to being surprised by the actual facts of nature. The doctrine of creation logically supports the inductive method of natural science that always makes room for endless new fields of research.

SUGGESTIONS FOR FURTHER READING AND STUDY

Davies, Paul. *The Mind of God: The Scientific Basis for a Rational World.* New York: Simon & Schuster, 1993.

Moltmann, Jürgen. *God in Creation: A New Theology of Creation and the Spirit of God.* Translated by Margaret Kohl. San Francisco: Harper & Row, 1985.

Rees, Martin. *Our Cosmic Habitat.* Princeton: Princeton University Press, 2001.

Russell, Robert John, Nancey Murphy, and C. J. Isham, eds. *Quantum Cosmology and the Laws of Nature.* Notre Dame: Vatican Observatory and University of Notre Dame Press, 1997.

---

41. Ibid.

8

Life and the Spirit

A QUESTION ON THE MINDS of many scientists and philosophers today is how to account for the origin of living organisms. This issue cannot be a matter of indifference to a theology of nature wherein God is taken to be the author of life. But where can theology locate its own kind of explanation in relation to scientific accounts of life? Can science attribute the origin of life to physical and chemical processes without competing or conflicting with theology's ascribing the origin of life to the creative Spirit of God? In the intellectual world today, as I have been saying, there is a powerful temptation to settle for purely naturalistic accounts of everything. Scientific naturalism, the belief that the world available to scientific inquiry is all there is, takes the origin and functioning of life to be purely natural occurrences. To assign these to the special action of God, therefore, would render science pointless. Why bother to look for theological explanations of life, the naturalist asks, when science suffices?

Naturalism is the dominant belief system in scientific and philosophical circles today. There are, of course, different ways of understanding naturalism—such varieties as religious naturalism, soft naturalism, hard naturalism, and so forth. But in this book I am using the term "naturalism" in the way that most philosophers, scientists, and theologians do, namely, to designate the belief that nature, as made available to ordinary experience and scientific method, is literally *all* there is. Naturalism allows no room for the miraculous or the supernatural. Life and its origin, therefore, must be explained only in physical terms.[1] Theology's appeal to the Spirit of God as the source of life is useless conjecture.

Until not too long ago, naturalistic explanations seemed inadequate, especially in the life sciences. As recently as the early twentieth century, vitalistic

---

1. For this understanding of the term "naturalism," see Charley Hardwick, *Events of Grace: Naturalism, Existentialism, and Theology* (New York: Cambridge University Press, 1996). A more fully developed discussion of theology and the origin of life appears in my book *Is Nature Enough? Meaning and Truth in the Age of Science* (Cambridge: Cambridge University Press, 2006).

assumptions were still popular and in some quarters scientifically respectable. "Vitalism," from the Latin word *vita* (life), claims that a complete explanation of life must appeal at some point to a nonmaterial force that intervenes in nature and elevates lifeless matter to the status of life. Henri Bergson, the most famous of modern vitalists, referred to this supramaterial force as a "vital impetus" (*élan vital*) whose function is suggestive of the work of the Holy Spirit.[2] According to vitalists, since there is something mysteriously supernatural about life, science can say very little about what it really is. The mere specifying of physical and chemical causes is not nearly enough to explain either the origin or the essence of life. Scientists in the not too distant past, many of whom who were influenced by vitalism, hardly dared to stray into an area of inquiry that borders so intimately on the spiritual as the realm of life seems to do. However, as we shall see in a moment, the scientific understanding of life is quite different today.

Until several centuries ago, human thought was hardly ever predisposed to embrace naturalism. Prior to the emergence of modern physics, geology, evolutionary biology, and Big Bang cosmology, the world was pictured as a fixed hierarchy of distinct levels of being, most of which were thought to have a spiritual quality inherent in them. If matter can be represented by the letter m, then plant life is $m+x$, where x is the intangible quality that a vitalizing principle adds to the chemical makeup of flowers, grass, and trees. Next, animal life and the beginning of consciousness can be represented as $m+x+y$, where y stands for the elementary sentience and even consciousness that nonhuman organisms possess. Finally humans, with their capacity for reflective self-awareness, can be coded as $m+x+y+z$. Until recently the indefinable x, y, and z dimensions seemed far beyond the range of scientific comprehension.[3]

A major implication of this hierarchical cosmology is that the higher levels cannot be reduced to the lower. There is something distinctive about each level, an ontological discontinuity that renders the higher levels more valuable and more real, though also more elusive, than the lower. Consequently, life cannot be explained solely in terms of the hard sciences, physics and chemistry. The latter can understand much (though not everything) about the lowest level, that of lifeless matter, but they are not equipped cognitionally to capture the nonphysical qualities in life, consciousness, and self-awareness. Even in the age of science a hierarchical view of nature continues far and wide

2. Henri Bergson, *Creative Evolution*, trans. Arthur Mitchell (Lanham, Md.: University Press of America, 1983), 88-97.

3. E. F. Schumacher, *A Guide for the Perplexed* (New York: Harper Colophon Books, 1978), 18ff.

to shape the ethical and legal sensibilities of human societies. People instinctively continue to show more reverence for living, sentient, and conscious beings than they do for rocks. At the same time, however, in modern times there appears to be less reverence for life than ever before in human history. Today this devaluation of life is sanctioned by materialist interpretations of nature that enshrine lifeless matter as the ground of all being, thus making life and mind merely derivative or secondary.

THE ORIGINS OF NATURALISM

How, then, did the naturalist victory over vitalism come about? The story is complicated and no version captures every nuance, but the profound Jewish thinker Hans Jonas provides an illuminating interpretation of this momentous drama, one that is of considerable importance to a theology of nature.[4] Jonas begins by observing that prior to the scientific revolution most people clung tightly to a *panvitalistic* understanding of the world. Panvitalism is the belief that *all* of reality is living. In the thought world of our remote ancestors not only flora and fauna but also the sun, stars, weather patterns, rivers, and landscapes pulsed with life. Nothing could be real without also being somehow alive.

But if life is the fundamental reality to the panvitalist, then what is death? How can death be real if everything real is supposed to be alive? Imagine for a moment that you are living in a panvitalist's world and that an animal or a person in your tribe has died. As you examine the dead body lying there in front of you, what are your thoughts? According to Jonas's account, you would be extremely puzzled, since the lifelessness of the corpse simply does not fit the overriding assumption that everything real is alive.

Jonas speculates that for most of our ancestors life was the norm and death the unintelligible exception. And so, in order to safeguard their panvitalistic worldview, people of the remote past instinctively came up with the idea of the *soul*, an imperishable animating principle that inhabits each organism. The soul is taken to be an intangible subjective center that still lives on mysteriously even though the body that it had previously animated is now motionless. To the panvitalist the essential core of living beings, animals along with humans, still lives on somewhere indefinitely. Souls may come back at times to haunt or to console people, but in either case they are taken to be more real

---

4. Hans Jonas, *The Phenomenon of Life* (New York: Harper & Row, 1966).

than the bodies they had enlivened. Thus, belief in the existence of souls allows the panvitalist to hold onto the presumption that life is more real than death.

Belief in the existence of the soul, it goes without saying, has relieved the anxiety about death for millions upon millions of people throughout the ages. But in modern times, Jonas proposes, the soul's immortality has been secured at the expense of the death of nature. For once the era of empirical science arrived, a sense of the soul's independence of the material world ironically permitted a reverse assumption to take hold, namely, that the material world considered by itself is soulless and lifeless. An ancient religious tendency to split reality into souls, on the one hand, and soulless materiality, on the other, could not help but prepare the way for the influential mind/matter dualism of René Descartes (1596-1650). For Descartes there are two very different kinds of being: thinking substance (mind) and extended substance (matter). This dualistic worldview has had the effect of exorcising anything spiritual, lifelike, or mental from what appears to be the inherently lifeless and mindless world of matter.

Consequently, in modern times the sense of a sharp split between mind and matter has settled into philosophy, theology, spirituality, and science. It still hovers over the world of contemporary thought, and vestiges of animism and dualism still live on in our religions. However, for our purposes here it is most important to point out that scientific naturalism, which has generally tended to be materialistic, is an outcome of the dualistic expulsion of any traces of a life principle from the physical world. Furthermore, the idea of an essentially lifeless and mindless realm of matter has come to serve as the philosophical foundation of modern scientific thought.[5] In scientific research, life and mind, accordingly, have come to be thought of as mere derivatives of a lifeless and mindless material substrate.

According to Jonas it was the incipient dualism of premodern thought that prepared the way for modern thought's radical expulsion of mentality and vitality from the realm of the physical. It is dualism that made possible the idea that the material universe is essentially and pervasively dead, a worldview that was not widely entertained until rather recently in the history of thought. So after the birth of modern science, and especially in the wake of astronomical disclosures of the vast domain of lifelessness in outer space, the physical universe has come to be thought of as essentially devoid of life except perhaps

---

5. E. A. Burtt, *The Metaphysical Foundations of Modern Science* (Garden City, N.Y.: Doubleday Anchor Books, 1954).

for a small patch here and there. Deadness has increasingly become the norm, that is, the natural state of things, and life is now the unintelligible exception that needs to be explained in terms of what is lifeless.[6]

Along with Jonas, theologian Paul Tillich refers evocatively to this modern naturalistic view as an "ontology of death." It is deadness, not aliveness, that now claims the status of being most real, especially in circles of thought influenced by scientific materialism.[7] In a most fascinating way, the ancient pan-vitalist problematic has been turned upside down. For our ancestors, life was the norm, and death the unintelligible exception that begged for explanation. Today, at least in much of the intellectual world, death is the norm, and life the unintelligible exception. Scientific research programs the world over now embrace the assumption that the universe is essentially dead. So they are driven to explain in terms of lifeless bits of matter how something as wondrous as life could possibly have sprouted out of the unpromising soil of an inherently dead cosmos.

In Jonas's own words:

> From the physical sciences there spread over the conception of all existence an ontology whose model entity is pure matter, stripped of all features of life. What at the animistic stage was not even discovered has in the meantime conquered the vision of reality, entirely ousting its counterpart. The tremendously enlarged universe of modern cosmology is conceived as a field of inanimate masses and forces which operate according to the laws of inertia and of quantitative distribution in space. This denuded substratum of all reality could only be arrived at through a progressive expurgation of vital features from the physical record and through strict abstention from projecting into its image our own felt aliveness.[8]

As a result, for modern scientific consciousness, "it is the existence of life within a mechanical universe which now calls for an explanation, and explanation has to be in terms of the lifeless."[9]

The explanation of the living in terms of what is dead continues to be the methodological objective of much scientific effort. Since life is said to be composed of dead matter, the truly explanatory sciences must therefore be chemistry and physics. The traditional lines of demarcation that located humans, animals, plants, and minerals on separate ontological levels have dissolved,

---

6. Jonas, *Phenomenon of Life*, 9-10.
7. Paul Tillich, *Systematic Theology*, 3 vols. (Chicago: University of Chicago Press, 1963), 3:19.
8. Jonas, *Phenomenon of Life*, 9-10.
9. Ibid.

and lifeless "matter" has assumed the status of being the ultimate ground and explanation of everything else. In keeping with developments in physics, some scientists and philosophers have recently begun to question the crude materialist reductionism dominant in contemporary attempts to account for life. But the philosophical foundation of most contemporary biology and neuroscience is still predominantly an ontology (i.e., an understanding of being) in which lifelessness is held to be more real and more intelligible than life.

Obviously, we still *feel* ourselves to be alive, but modern scientific naturalism has instructed us not to project our subjective feelings onto the originally lifeless and mindless world "out there." A notable illustration of this prohibition is the French biochemist Jacques Monod's book *Chance and Necessity*, a work that appeared in the late 1960s and rapidly became a materialist classic.[10] It still stands today as a monument to the Cartesian expurgation of life and mind from the physical record. For Monod the living cell is nothing but a material mechanism, so the key to understanding life is to learn the physics and chemistry underlying cellular activity. Any other way of understanding nature would be a violation of what Monod calls "the postulate of objectivity," an ordinance that forbids scientists to entertain vitalistic or animistic hypotheses.[11]

From Jonas's point of view, Monod's materialist manifesto is emblematic of the fact that for scientific naturalism "the lifeless has become the knowable par excellence and is for that reason also considered the true and only foundation of reality." Lifelessness has come to be understood as "the 'natural' as well as the original state of things."[12] The reader may wish to recall here the thirty volumes of the fourteen-billion-year-old cosmic story I presented earlier, where the first twenty-two books are devoid of life. That the universe lacked living organisms until only 3.8 billion years ago appears to support the view that dead matter is the mother of all things.

It is not surprising, then, that the impression of a pervasive deadness and mindlessness as the ground state of the universe continues, though not without significant exceptions, to shape the current agenda of much scientific research. Francis Crick's book *Of Molecules and Men*, for example, asserts that "the ultimate aim of the modern movement in biology is to explain all of life in terms of physics and chemistry."[13] It is hard to imagine a more radical

---

10. Jacques Monod, *Chance and Necessity: An Essay on the Natural Philosophy of Modern Biology*, trans. Austryn Wainhouse (New York: Knopf, 1971).

11. Ibid., 175-80.

12. Jonas, *Phenomenon of Life*, 9-10.

13. Francis H. C. Crick, *Of Molecules and Men* (Seattle: University of Washington Press, 1966), 10.

inversion of the hierarchical cosmology embraced by many of the world's religions, including Christianity. According to Crick, a heroic figure in the annals of science, the whole ladder of living and thinking beings can now be understood fully in terms of the lowest level of the classic hierarchy of being. Henceforth, as Crick's associate James Watson adds, "life will be completely understood in terms of the coordinated interactions of large and small molecules."[14] And according to the philosopher Daniel Dennett, not only life but also mind is now reducible to lifeless matter:

> There is only one sort of stuff, namely matter—the physical stuff of physics, chemistry, and physiology—and the mind is somehow nothing but a physical phenomenon. In short, the mind is the brain. According to the materialists we can (in principle!) account for every mental phenomenon using the same physical principles, laws and raw materials that suffice to explain radioactivity, continental drift, photosynthesis, reproduction, nutrition and growth.[15]

Such a claim may seem extreme, but the esteem in which Dennett is held by scientists and philosophers today is indicative of the overall physicalist leaning of much academic thought.

THE COSMIC EXTENSION OF DEADNESS

The ancient panvitalist problematic of how to account for the anomalous fact of death if everything is throbbing with life has given way to the contemporary riddle of how to explain life if the world is normally and habitually dead. Cartesian dualism, Jonas writes, was an intermediate stage in this transition, and in its own flirtations with dualism, theology has perhaps unwittingly helped pave the way for an ontology of death as the counterpart to the world of life and the soul. By segregating the soul from the natural world, certain trends in traditional theology have helped to bring about the fiction of an essentially lifeless and mindless cosmos. Centuries of theologically sponsored dualism, having implicitly drained life and mind from nature, have thus passed on to modern thought the makings of the very ontology of death that now renders belief that God is the author of life nearly impossible for countless educated people to accept.

---

14. J. D. Watson, *The Molecular Biology of the Gene* (New York: W. A. Benjamin, 1965), 67.
15. Daniel C. Dennett, *Consciousness Explained* (New York: Little, Brown, 1991), 33.

The actual discoveries of modern and recent science have added an air of empirical credibility to the ontology of death that had already been put in place by dualism and its materialist offshoots. For example, the awakening by science to deep geological and cosmic time, according to which the universe is devoid of living beings throughout most of its history, only reinforces the naturalist's suspicion that life was never intended to happen in the first place. Likewise the astronomical impression that life, quantitatively speaking, makes up only an infinitesimal amount of the immense cosmic mass makes life on earth seem a rare exception to the lifeless norm.

Even if maverick astrobiologists are correct in conjecturing that there are extraterrestrial precincts of life in the wider Big Bang cosmos, the prevalent feeling among scientists, at least until quite recently, has been that life fits only uneasily into the pervasive deadness and silence of nature. And even if some astrophysicists now consider life to be an inevitable outcome of the way the Big Bang cosmos is set up, there is still a suspicion that life and thought are not essential to the cosmos as such. Indeed some scientific naturalists today are taking imaginative extra steps to protect their assumption that nature is essentially lifeless. For example, the renowned cosmologist Martin Rees is willing to admit that the emergence of life in the Big Bang universe is nearly inevitable, given its initial physical conditions and fundamental constants. But the possible existence of a multiverse can enlarge the natural world so lavishly that the fact of life in our own universe remains anomalous and negligible when situated against the backdrop of a cosmic environment consisting of mostly lifeless universes. The existence of any universe endowed with biophilic traits (such as our own) may seem at first to challenge the ontology of death, but scientific speculation can still satisfy the naturalistic inclination to make the *whole* of nature appear essentially lifeless. All it has to do is multiply universes with unrestrained imagination in such a way that almost all of them will be construed as devoid of life.

Just suppose (since direct evidence is lacking) that there are countless many universes that exist as the larger setting for our own. Then you can save the idea of a persisting cosmic deadness by viewing our own life-entertaining Big Bang cosmos as a fleeting exception in an immensely larger plurality of universes, most of which would be structurally indifferent to life. Thus, as a statistical whole the enlarged natural world would still remain foundationally lifeless. Although it is not his stated intention to do so, in effect Martin Rees, along with other multiverse enthusiasts, is looking for theoretical room to let an *intrinsic* lifelessness and impersonality stand firm as the natural state of being. In order for life and mind to retain their bearing as alien exceptions to the *true* nature of things, multiverse conjectures allow our own peculiarly life-bearing universe to be an aberration in an immense company of mostly still-born cosmic experiments. This speculation, which so far can claim to be based on no evidence whatsoever, is appealing to the naturalist since it allows the idea of an intrinsically dead cosmos to remain intact.[16]

In the following chapter I shall try to draw out some of the more momentous implications of the modern ontology of death for theology's encounter with science. But for now I want to make it clear once again that Christian theology has no reason to object in principle to the idea of a multiverse as such. This fascinating idea is perfectly consistent with the extravagance of divine creativity as well as with certain interpretations of quantum physics and what has come to be known as string theory. What I am highlighting here is the extent to which the naturalist way of thinking will go, in the absence of any scientific evidence, in order to preserve the belief that lifelessness is the most natural and intelligible state of being.

ROOM FOR THEOLOGY?

So, in view of the intellectual respectability that physicalist naturalism still enjoys today, can there be a meaningful place anywhere for a theological explanation of life? How does one conceive of the role of the *Creator Spiritus*? Traditionally, religious scholars made a distinction between primary and secondary causes: God is the ultimate or primary explanation of everything, but the Creator works through secondary or proximate natural causes. Science can deal with secondary or natural causes, but it has no access to the primary supernatural cause of everything. Many theologians still follow this distinction, arguing that a scientific analysis of the chemical causes of life allows ample room for understanding God as primary cause. However, I believe the conversation of theology with science requires a slightly different approach today.

What I shall propose here is that a theology of nature must first make a case for the plausibility of "layered explanation" before it begins trying to show how life may be explained simultaneously by both science and theology. By layered explanation I mean to point out that most things in our experience admit of more than one level of explanation. Even the simplest occurrences in our experience involve a plurality of such levels, so before going any further it may be appropriate to demonstrate how, at least in principle, there may be

---

16. Martin Rees, *Our Cosmic Habitat* (Princeton: Princeton University Press, 2001); see also Lee Smolin, *The Life of the Cosmos* (New York: Oxford University Press, 1997).

room logically for both scientific and theological levels of understanding natural phenomena. Scientific naturalism, of course, will protest that there is no need for explanatory pluralism, and so it leans typically toward "explanatory monism." That is, it declares that if a physical explanation of life is available then a theological explanation is unnecessary. Layered explanation, however, allows space for both theological and scientific understanding, for both the Spirit of God and natural processes. There is no need to assume that there is any real conflict or competition between them.

Let me provide here a simple example of what I mean by layered explanation, or as it may also be called, "explanatory pluralism." Suppose there is a pot of water boiling on your stove.[17] A friend comes by and asks you why it is boiling. You may answer your friend's question by saying that it's boiling because the molecules of water are escaping as the pot heats up. This is a perfectly good explanation, but it does not rule out others. You may also tell your friend that the pot is boiling because you turned the stove on, also a perfectly good explanation, but one that allows for even deeper explanation. You may respond, third, that the pot is boiling because you want to brew some tea. Each of the three explanations may be offered without any one of them competing with or ruling out the others. Each explanation is only an abstract selection from the complex totality of causal factors involved in making the pot boil. The point is that rich understanding of anything requires our taking into account a plurality of explanatory factors.

I offer this simple example as a way of showing that, at least in principle, there is room for more than one level of understanding almost anything. Accordingly, there may be room for both theological and scientific explanations of any phenomenon in the natural world. As regards the question of life's origin, to return to this chapter's topic, scientific explanations, no matter how detailed they become, do not logically conflict or compete with genuinely theological explanations. It would make no sense for me to tell my friend that the pot is boiling because of molecular activity *rather than* because I want some tea. Likewise I do not reply that the pot of water is boiling because I want tea *rather than* because I turned the stove on. Analogously, no matter how brilliant or convincing a scientific theory of life's origin may be, I am not obliged to conclude, as materialist scientists and philosophers do, that life came about on our planet because of a specifiable concatenation of physical

17. This example is provided by John Polkinghorne in *Quarks, Chaos and Christianity: Questions to Science and Religion* (New York: Crossroad, 2000), but I am taking considerable liberties with it here.

events *rather than* because of divine goodness and generosity. There is room for multiple levels of explanation.

In any instance of layered explanation the different levels of understanding do not compete, nor do they have to be correlated point by point, with one another. For example, as I examine the physics of water and steam, I do not expect to see "I want tea" inscribed in the molecules of water. Nor do I need to worry about the physics and chemistry of water or steam while I am forming a mental purpose to drink some tea. Completely different, though non-competing, causal levels can be operative in the production of a single event, and I need to keep alive a sense of this plurality if I am to avoid the fallacy of "reductionism."

A good definition of reductionism is "the suppression of layered explanation." But here it is necessary to make a distinction between scientific reduction and the logical fallacy known as reductionism. Reduction is the legitimate scientific method of breaking down comprehensive wholes into their component parts. There is no need to be critical of this important way of understanding phenomena, including living organisms. *Reductionism* on the other hand is the arbitrary declaration that there can be only one level of understanding. It is the refusal to allow for a plurality of explanatory levels. As such, reductionism is a fallacy that can be committed no less by religious believers than by scientists, for example, when a creationist claims that it was God *rather than* chemical processes that accounts for the origin of life.

RELATING SCIENCE TO THEOLOGY

The elusiveness of theological explanation, of course, is annoying to the naturalist whose own brand of reductionism declares that everything real must be laid open with the same degree of evidential and mathematical clarity that science tries to provide. The modern naturalistic ideal has usually also been a Cartesian one, namely, that clear and distinct ideas are essential to fundamental explanation.[18] However, this amounts to a logical confusion of what is

---

18. See Alfred North Whitehead, *Process and Reality,* corrected ed., ed. David Ray Griffin and Donald W. Sherburne (New York: Free Press, 1978), 162. "It must be remembered that clearness in consciousness is no evidence for primitiveness in the genetic process: the opposite doctrine is more nearly true" (p. 173). We should seek clarity, but then we should mistrust it, since, as Whitehead argues, clarity comes about only as the result of our leaving out most of the tangled web of events that make up the concrete world.

elemental with what is *fundamental.*[19] Theology is in search of fundamental or ultimate explanation, and it can provide this only in the language of analogy, symbol, and metaphor. Furthermore, to Christian theology the real meaning of things remains to be fully revealed in God's future. Hence theology must not emulate the mathematical precision of natural science, nor need it apologize for the lack of clarity in its use of symbolic discourse. While the elemental can be grasped with some degree of clarity, that which is truly fundamental resists such easy capture. It grasps us more than we grasp it.

Consequently, I propose that divine influence stands in relation to the natural world, including such events as the origin of life, analogously to the way in which "I want tea" stands in relation to the molecular commotion in the boiling pot of water on my stove. Even the most painstaking examination of the molecular activity in the pot of boiling water is not going to reveal, at that level of analysis, the "I want tea" that contextualizes and motivates the whole scene involving the boiling of water. But the fact that I want tea, even though it does not show up at the level of the elemental water molecules, is still the "ultimate" explanation of the water's boiling. The lesson one may draw from this very simple example is that even the most detailed scientific examination of the complex sequence of physical and chemical events that led up to the emergence of the first living cell, says nothing about any ultimate reason why life came about in the universe at all. Furthermore, there is no danger that as scientific studies of the origin of life become more precise, accurate, and convincing, theological explanation will become weaker and less relevant. Layered explanation forbids such senseless competition.

Let us look at some other examples of layered explanation, just to reinforce the habit of mind I am advocating. As you read this page your mind is active. But how are you to explain why at this moment you are thinking and raising questions? One very good explanation is that you are thinking because the neurons in your brain are firing, the synapses are connecting, and your parietal lobes are being activated—all the fascinating things that you can learn about from scientific studies of the brain and nervous system. Again, the neuroscientific accounts are reasonable and essential, and they should be taken as far as research allows. But without in any way belittling scientific explanations of mind you can also answer the question "why are you thinking?" by declaring that it is because of your *desire to understand* what is going on in the world. Distinct levels of understanding are simultaneously applicable here also.

---

19. To confuse abstractions with concrete reality is a logical fallacy, the "fallacy of misplaced concreteness." Whitehead thinks that much of modern thought is based on this fallacy. See Alfred North Whitehead, *Science and the Modern World* (New York: Free Press, 1925), 54-55, 51-57, 58-59.

Predictably the naturalist, especially the eliminative materialist or the hard reductionist, will try to map your second explanation completely onto the first. It is the intoxicating dream of providing such a simplification that gets materialist naturalists up every morning. But their bold expectations can never be fulfilled because of the elementary fact that the first-person perspective you take in your second explanation cannot be integrated smoothly into the third-person perspective of the neuroscientist. The gap between first-person, subjective experience, on the one hand, and the objectifying approach of scientific method, on the other, is known today as "the hard problem" in cognitive science, and philosophers of mind are becoming increasingly sensitive to it.[20] But if more scientists and philosophers would become familiar with the possibility of layered explanation it would become obvious that multiple levels of explanation are needed, each irreducible to the others.

To go just a bit deeper with this example, if you really want to understand why you are thinking, at some level a third answer is required: you are thinking *because reality is intelligible.* If reality were not intelligible, your mind would not be working at all. A comparison to ocular vision is appropriate here. Without an environment that bathes the world in light there would never have been any gradual awakening of vision or any evolution of eyes in natural history. Analogously, a necessary environment for the emergence of mind in evolution is that the universe has always been intelligible, long before minds came along. But then a deeper question arises: Why is the universe intelligible at all? In response it seems quite reasonable to make a place for theological explanation in any richly layered understanding of human intelligence. Theology is permitted to claim that intelligence arose in the history of nature *ultimately* because the universe is grounded in an eternal, creative Principle of Intelligibility. Such an explanation would in no way compete with biological accounts of the emergence of mind in evolution or with neuroscientific explanations of how the mind works. Each level can be pushed to its limits without rivaling or causing any conflict with the others. That is, you don't have to say that you are thinking because your neurons are firing *rather than* because you are trying to understand, or *rather than* because the universe is intelligible. And you don't have to insist—as do Darwinian materialists—that natural selection *rather than* divine wisdom is the ultimate cause of intelligence. Different explanatory levels can exist in harmony, side by side, and one need not be collapsed into another.

---

20. David Chalmers, "Facing Up to the Problem of Consciousness," *Journal of Consciousness Studies* 2 (1995): 200-219.

The theologian, therefore, can be fully content to let science account for mental as well as living beings in a physical or elemental way as long as room remains also for more fundamental explanation. Contrary to the allowance for layered explanation, however, *reductionism* is not a method of knowing at all, but a suppression of knowledge that tries to replace what is fundamental with what is elemental. Rather than being the expression of a humble desire to know, reductionism is the manifestation of a *will* to control, seeking by fiat to force all possible explanations into a single manageable level where the obsessive need for immediate clarity rules out rich understanding.

The materialist type of reductionistic suppression of layered explanation is an expression of a will to mastery rather than a cognitionally fertile openness to truth. But theologians and religious people can be no less reductionistic when they thoughtlessly declare that life came about on earth because of God's creativity *rather than* because of chemical processes, or when they insist that it was divine action *rather than* evolutionary processes that brought about distinct species. In both naturalistic and religious types of reductionism the questionable assumption is that there must be only one level of explanation and that divine creativity is somehow in competition with natural causes. The way to avoid apparent conflicts between science and theology, therefore, is to allow generously for thickly layered explanation.

EXPLAINING LIFE

Let us come back, then, to the intriguing question: Why did life appear on earth? Today one can notice immediately that even within the world of science alone an explanatory pluralism is already at work in dealing with this fascinating puzzle. Physicists, for example, try to explain the birth of life in terms of thermodynamics or the self-organizing tendencies of matter. Chemists—and here I am obviously oversimplifying—explain the origin of life in terms of the way carbon bonds with other atoms and how complex organic molecules can combine in cellular mechanisms. Biochemists speculate about the role of early RNA or protein replication. Planetary scientists, geologists, and ecologists also make valuable contributions to a scientific understanding of life's origin. And recently astrophysicists have entered the discussion by demonstrating that inquirers cannot even begin to explain how and why life came about on earth without going all the way back to the opening microseconds of the Big Bang universe's existence. If life was to come about eventually in our universe, several of its physical features, such as the force of gravity or the expansion rate of the universe, had to possess just the right numerical values from the very beginning. In order to make scientific sense of the origin of life locally it is necessary to take into account certain traits that are constitutive of the *whole universe*.[21]

When it comes to understanding life even in terms of science, therefore, one can now discern a thicker layering of explanation than ever before. If layered explanation is legitimate, then it would not be illogical for theology to suppose that there is also room in principle for a fundamental kind of explanation that would complement rather than contradict the various sciences. Without competing with science, theology may propose that life came about and diversified on earth because of the creative and vitalizing power of God's Spirit.[22] Just as my wanting a cup of tea does not compete with a physical understanding of the water boiling on my stove, so also it is logically plausible to maintain that life arose on earth because of the creative influence of God, without denying that chemical and astrophysical factors are also at work. Naturalists and creationists will still insist that we have to make a choice between the two accounts, but there is no logical reason to do so.

The Analogy of Information

God's creation and energizing of life in the universe can occur so quietly and unobtrusively that the laws of chemistry and physics are in no way violated. Another analogy may help us understand how God's Spirit can be deeply active in the vitalizing of the universe without violating the laws of chemistry and physics in the slightest way, or without requiring an unnecessary philosophical vitalism. Imagine that you are doodling with your pen on a scrap of paper. Then, without lifting your pen from the paper you abruptly start writing a meaningful sentence, using letters of the alphabet. From a purely chemical point of view—let us say, from the point of view of the chemistry that bonds ink to paper—your methodical writing looks no different from the scribbling. If the level at which you wish to explain the writing on the page is that of chemistry, you are not going to see anything different in your deliberate writing from what you see going on in the aimless scrawl. The chemical laws that allow ink to appear on paper are the same in both instances. From a certain point of view, therefore, you can say that it is all "just chemistry." But anyone who knows how to read will immediately see something going on in

---

21. Rees, *Our Cosmic Habitat.*

22. As Jürgen Moltmann perceptively notes, it is the function of God's Spirit not so much to spiritualize as to *vitalize* the world (*The Spirit of Life: A Universal Affirmation*, trans. Margaret Kohl [Minneapolis: Fortress, 1992], 74, 83-98).

your deliberately contrived sentence that a purely chemical approach will miss. A reader, anyone who has acquired the skill to recognize a higher organizational level, will observe that there are letters of a code arranged in a *specific sequence*. There is something informational going on in the meaningful arrangement of letters that a chemical analysis of the page cannot apprehend.[23]

But let us look even closer. You will notice that while you were writing the meaningful informational sequence you violated none of the physical or chemical laws that had been functioning deterministically while you were merely scribbling. When you are writing the sentence there is no miraculous suspension of the chemical laws that bond ink to paper. Meaningful information came onto the page without your interrupting in the slightest way the physical or chemical laws that allow you to apply ink to paper. Something powerfully new and significant happened on the paper without being expressible in terms of chemistry.

Analogously, the Spirit of Life may be powerfully at work in creation without ever being noticed by the natural sciences and ordinary awareness. When life comes into the cosmos, the continuum of physical causes and effects is never broken. The Author of Life may act to enliven the universe without ever showing up in the sphere of scientific inquiry. Information came onto your written page without interrupting anything—physically and chemically speaking—but it would not have been noticed at all if you had not been able to read. Indeed it employed, rather than violated, the chemistry of ink and paper. Yet everything was changed dramatically. So the Spirit of Life can make the world radically new without in any way violating the timeless laws that bond carbon to oxygen, nitrogen, hydrogen, and so on.

Information makes all the difference in the world, but it does not disturb anything, physically speaking. Information "works" its wonders by way of a gentle and nonintrusive effectiveness. The influence of God's Spirit on the natural world may be *analogous* to the nonintrusive effectiveness of information. Divine action should never show up as such at any of the levels familiar to scientific inquiry. For that reason it is possible to hold that there is no real competition between scientific and theological explanations of the origin of life. It would be a cheap and crude theology indeed that would try to place divine action at any of the levels of explanation employed by the various sciences. Conversely, it is unwarranted for the naturalist to rule out divine influ-

---

23. Here I am adapting an example provided by Michael Polanyi. See his *Knowing and Being,* ed. Marjorie Grene (Chicago: University of Chicago Press, 1969), 22-39, 229.

ence in nature just because no "evidence" for it appears at the various levels specifiable by scientific analysis.

An Essentially Lifeless Universe?

At one time the universe seemed completely alive, but in terms of the perspective of modern science, as Hans Jonas says, lifelessness has now become the most natural and intelligible state of being. Science as such, of course, is not itself responsible for the death of nature. Science is a method of inquiry, not a worldview, and so it is not fair to accuse it of perpetrating an ontology of death. As a method of inquiry that abstracts from qualities, vitality, and subjective feelings, science is simply not programmed to penetrate the mystery of life. Rather, it is scientism and scientific materialism that have squeezed the life out of nature. They have done so by first denying that subjectivity, both human and nonhuman, is really a part of nature. Then, after they have surgically removed subjectivity from the tissue of nature, it is only a small second step to devitalizing it also.

The background assumption of much contemporary thought, as noted above, is that being or reality is naturally lifeless (and mindless), and this helps explain why research on the origin of life fills up the careers of so many scientists today. Tacitly driving origin-of-life studies, as well as the nascent interest in astrobiology, is the intriguing question of how something so apparently "unnatural" as life could ever have come to birth out of the more "natural" and intelligible state of nonlife as exposed by the physical sciences. How could anything that differs so radically as living (and thinking) organisms do from the world's "natural" deadness ever have emerged without magic, miracle, or supernatural assistance from the ground state of lifeless and mindless "matter"? Modern scientific naturalists (many of whom are happy to be labeled materialists or physicalists) understand the essential deadness of nature to be an unbreakable continuum into which the apparently remarkable facts of life and mind must eventually be resolved by the power of scientific analysis in order to be rendered fully intelligible. In this way naturalists hope to reach their goal of leveling what seems initially remarkable down to what is really quite unremarkable.

Meanwhile, their dualistic religious opponents often ironically also embrace the same ontology of death as the most natural and intelligible state of terrestrial occurrences. Their view of nature is tolerated theologically because it seems to allow that special interruptive acts of God will stand out as all the more exceptional and supernatural. For example, creationists and intelligent design are willing to go along with their naturalist antagonists as far as understanding the nonliving world of nature is concerned.[24] The advantage they see in sharing this very modern view of nature is that it sets the stage for punctuating the blank everydayness of the world with dramatic divine displays. Without a preponderant backdrop of normal, natural deadness, religious opponents of contemporary neo-Darwinian biology fear that the power of God to create life might not be made visible at all. Accordingly, both the origin of life and the resurrection of Jesus must be understood as dramatic divine *interruptions* of what seems to be the quotidian paleness of being. Such theological expectations, however, are out of touch with the unobtrusive effectiveness of God's humility.

In their encounter with scientific naturalism, existentialist theologians of the twentieth century also sometimes conceded that nature is essentially lifeless, but they allowed that there also exists a more important realm of being distinct from nature, namely, that of freedom. In order to make sense of Jesus' resurrection Rudolf Bultmann, for example, located the mighty acts of God in the domain of human freedom, which he thought of as existing apart from nature. In this scientifically inaccessible arena, Bultmann proposed, all the religious drama necessary for a life of faith can take place without disturbing nature's normality.[25] This apologetic strategy seemed appealing, since it did not require scientists to give up the prevalent naturalistic assumption that nature is at bottom a lifeless set of mechanisms.

Unfortunately, the existentialist theological program places the core of our humanity outside of nature, and thus it fails to consider the possibility that Christianity's resurrection faith may have something to say about what is going on in the universe at large. But is there a reasonable alternative to either naturalism or religious dualism? Is there available a theological approach to nature that can take seriously the discoveries of science and at the same time situate the resurrection of Jesus—and our own hope to share in it—in some metaphysical space other than that permitted by dualism or scientific naturalism?

The emergence of life, naturalism proclaims, is a purely natural process, and

---

24. See Phillip E. Johnson, *The Wedge of Truth: Splitting the Foundations of Naturalism* (Downers Grove, Ill.: InterVarsity, 1999); Jonathan Wells, *Icons of Evolution: Science or Myth? Why Much of What We Teach about Evolution Is Wrong* (Washington, D.C.: Regnery, 2000); Michael J. Behe, *Darwin's Black Box: The Biochemical Challenge to Evolution* (New York: Free Press, 1996); William A. Dembski, *Intelligent Design: The Bridge between Science and Theology* (Downers Grove, Ill.: InterVarsity, 1999); for critiques of intelligent design theory, see John F. Haught, *God after Darwin: A Theology of Evolution* (Boulder, Colo.: Westview, 2000); idem, *Deeper Than Darwin: The Prospect for Religion in the Age of Evolution* (Boulder, Colo.: Westview, 2003); and Kenneth R. Miller, *Finding Darwin's God: A Scientist's Search for Common Ground between God and Evolution* (New York: Cliff Street Books, 1999).

25. Rudolf Bultmann, *The New Testament and Mythology and Other Basic Writings* (Minneapolis: Augsburg Fortress, 1984).

that means explicable fully in terms of the nonliving. Modern science, especially in combination with the recent discovery of deep cosmic time, has given the impression to many that our universe is *essentially* lifeless. Only the most unlikely concatenation of accidents (combined with routine physical processes), therefore, has allowed life to arise at all. By all accounts life seems unplanned. To the scientific naturalist, as we have seen, nature at bottom is an interminable deadness, and life appears only as a late, local, and apparently unplanned anomaly. The scientific discovery of deep cosmic time and the immensity of space has made it seem likely that life has only a precarious foothold in the universe.

Theology itself, as I noted above, is not without blame for bringing about this impression. The assumption that nature is essentially lifeless was sanctioned by modern religious thinkers who wished passionately to defend the notion of God's absolute sovereignty over nature. During the early years of modern science, the physicist and theologian Robert Boyle (1627-1691) assumed that the most appropriate way to defend the idea of divine transcendence and power is to conceive of nature itself as pure passivity. To him the essential deadness of a mechanistic universe did not seem out of keeping with a kind of theism in which God is in complete control. As far as Boyle and other mechanistic-minded theists were concerned, to attribute any kind of intrinsic creativity or spontaneity to the physical universe would diminish the sense of God's power over the universe, thereby placing the world in a competitive relationship with its Creator.[26] Assumptions such as Boyle's would later make it difficult for some theologians to accept the self-creativity of life as depicted by evolutionary biology.

The premodern hierarchies of nature that placed humans, animals, plants, and minerals on distinct levels in an essentially organic universe gave way in modern scientific naturalism to the assumption that pure "matter," stripped of any essential association with life, is the ultimate foundation of life and mind. And even though some scientists and philosophers have recently tried to move beyond simplistic forms of physicalism, they cannot help being constrained by the weight of intellectual habit to situate their reflections on life, evolution, and emergence within the governing framework of an ontology of lifelessness. As far as Christian theology is concerned, however, it is in no way contrary to scientific accounts of life's emergence to attribute the existence of life ultimately to the vitalizing power of the Holy Spirit.

---

26. See David Ray Griffin, *Religion and Scientific Naturalism: Overcoming the Conflicts* (Albany: State University of New York Press, 2000), 107-35.

SUGGESTIONS FOR FURTHER READING AND STUDY

Cobb, John B., Jr., and Charles Birch. *The Liberation of Life: From the Cell to the Community.* Cambridge: Cambridge University Press, 1988.

Jonas, Hans. *The Phenomenon of Life.* New York: Harper & Row, 1966.

Moltmann, Jürgen. *The Spirit of Life: A Universal Affirmation.* Translated by Margaret Kohl. Minneapolis: Fortress, 1992.

Tillich, Paul. *Systematic Theology.* 3 volumes. Chicago: University of Chicago Press, 1963. Part IV, "Life and the Spirit," 3:11-294.

9

Science, Death, and Resurrection

God did not make death, nor does he rejoice in the destruction of the living.

—Wisdom 1:13

I F THE NEWS OF JESUS' RESURRECTION from the dead was hard for his disciples and the first Christians to believe, it seems all the more so for those of us who live in the age of science. The earliest Christian witnesses experienced the resurrection as a completely surprising event, so rising from the dead would be even more startling to sensibilities shaped primarily by modern inductive methods of knowing. Science cannot make sense of unique events of any sort, let alone those so bereft of antecedents as the resurrection. Hence, even if it connects powerfully with the hope that persists in human hearts, Jesus' resurrection would appear scientifically to transcend all realistic expectations.

Science, as we have observed often, is most at home with generalizations based on observation of large numbers of similar events obeying invariant physical laws. It squirms anxiously in the presence of what is completely unprecedented, and it looks for ways to reduce what seems exceptional to what is already known. It is the nature of science to suppress the truly unique by fitting it into the universal. Indeed such an approach to understanding is a defining characteristic of modern intellectual culture. And so the difficulties that scientifically educated people have with the notion of resurrection stem in large measure from the nearly unshakable assumption that nature and history are simply not open to anything so impossibly new as Christ's victory over death.

It is important to add, however, that it is not science as such that renders Jesus' resurrection, along with the prospect of our sharing in his destiny, incredible. Science would simply pass over such an event without even noticing it. On the other hand, there can be no doubt that *scientific naturalism* stands in opposition to resurrection, whether that of Jesus or ourselves (in this chapter I am referring to both). According to scientific naturalism, all causes are natural causes, so there can be no events that cannot be explained exhaustively by scientific method. For that reason, the origin of life would have to be a purely natural event. One of naturalism's chief claims is that ideas about life after death "stand in the way of understanding our natures truthfully and locating what makes life meaningful in a nonillusory way."[1] However, as Owen Flanagan comments, once we have resigned ourselves to the gravity of naturalism, life does not have to be sad. The universe is pointless and death final, he declares, but human life can be meaningful and happy anyway. When we die our personalities disappear forever, but in the meantime we can live satisfying lives. Flanagan's is an interesting claim for many reasons, but here I want only to flag his assertion that belief in life after death, however one conceives of it, is "irrational." By this stern marker he means to emphasize that there is no scientific evidence that could conceivably support the expectation of resurrection or the subjective survival of death. Hence, reasonable people must not take the existence of the soul, immortality, or bodily resurrection seriously.

In the world of religious ideas, nothing, including belief in God, stands out as a violation of naturalistic assumptions more starkly than the expectation that the dead will rise again. It is simply not in the nature of things, naturalists claim, that resurrection could ever happen. On top of this, science has shown quite clearly that the direction of all physical processes, and of the universe as a whole, is toward final and irreversible disintegration. The entirety of life, of which human history is only a recent and precarious chapter, departs from the entropic cosmic decline only temporarily. Blank deadness will have the final word. And now that physics has linked human existence more tightly than ever to the natural world, the dreary outcome that awaits the whole universe seems to swallow up any hope that individual persons can escape it either.

What response can a Christian theology of nature make to this gloomy prognostication? At first it may be tempted to revert to the traditional idea of the soul's immortality as the simplest way out of the stated predicament. That is, it assumes that there is an immaterial part of us that, upon our dying, can be severed decisively from the cosmic plunge toward death by entropy. The second law of thermodynamics, with its news of a dying cosmos, may simply be ignored as theologically inconsequential. If the immortal human soul can break out of its material prison, the demise of the physical universe should have no bearing whatsoever on our hope for immortality.

---

1. Owen Flanagan, *The Problem of the Soul: Two Visions of Mind and How to Reconcile Them* (New York: Basic Books, 2002), 167-68.

This approach to the problem of death appeals to many religious believers as well as to those who adopt what I earlier called the "contrast" approach to science and theology. It accepts the judgments of science and sees no conflict between science and religion, but it tries to protect theology from having to undergo any major changes in the light of new scientific ideas. To many of its proponents metaphysical dualism—an ancient worldview that allows for a final separation of mind, soul, and spirit from matter—may still seem to be the most efficient way to accommodate both the requirements of science and the hopes of religion.

Unfortunately, however, from the point of view of Christian faith such dualism is objectionable. Christians believe in bodily resurrection, and bodies are inseparable from the material universe. In some sense, therefore, resurrection, if it is not an irrational belief, must be the destiny of the entire universe, not simply of perishable human lives. So Christian theology, today more than ever, must find a way to connect the whole cosmic story to that of Jesus Christ "in whom all things consist" (Col 1:17). Ironically, such a theology will agree immediately with the naturalist on the inseparability of human and cosmic destiny. The difference is that the naturalist rejects all hope for resurrection because the universe disallows it, whereas the Christian must have hope for the whole universe because the resurrection of Jesus demands it.

In any case, scientifically informed theology must now start with the premise that humans and the cosmos are forever interlocked. Unlike the contrast approach, which sees little relevance in science for theology, the contact approach that I am endorsing throughout this book is challenged by geology, cosmology, and biology to widen our awareness of what is saved by Christ. The redemptive meaning of his death and resurrection points to much more than the salvaging of souls from the universe. Rather, a resurrection faith implies that in Jesus' destiny that of the entire universe is at stake. If Christ is not risen from the dead, then vain is our faith and naturalism wins. But if he is risen indeed, then theological consistency requires that we bring the entire universe into the sweep of what is destined for redemption. A bit later on I will set forth a proposal as to how theology may conceptually unfold this idea in the face of cosmology's certainty that far down the road the universe itself will fade away.

HOW CAN WE UNDERSTAND "RESURRECTION"?

Owen Flanagan, cited above, admits that most people still believe that the human soul or self will live on after death. But to him such obsolete beliefs are an annoying impediment to the spread of naturalism, a dominant strain of contemporary thought. There are different kinds of naturalism, but Flanagan's is uncompromisingly materialistic. Ever since the seventeenth century some of the most influential scientific materialists have sponsored the belief that matter is all that is real.[2] Consciousness itself, according to materialists, will eventually dissolve into the mindlessness of matter. Disturbed by such a prospect, the great American psychologist and philosopher William James provides a candid summation of what materialist naturalism logically entails as far as natural and human achievements are concerned:

> That is the sting of it, that in the vast driftings of the cosmic weather, though many a jewelled shore appears, and many an enchanted cloud-bank floats away, long lingering ere it be dissolved—even as our world now lingers for our joy—yet when these transient products are gone, nothing, absolutely *nothing* remains, to represent those particular qualities, those elements of preciousness which they may have enshrined. Dead and gone are they, gone utterly from the very sphere and room of being. Without an echo; without a memory; without an influence on aught that may come after, to make it care for similar ideals. This utter final wreck and tragedy is of the essence of scientific materialism as at present understood.[3]

The vast majority of the earth's inhabitants today still share the sentiments of James rather than Flanagan. They would consider the final extinguishing of minds and persons as well as the utter obliteration of all the impressive ethical and aesthetic achievements of humanity throughout history to be the greatest of evils. They deny, either explicitly or implicitly, that everything ends up in absolute nothingness. Of course, their instinctive revulsion is no proof that they are right, but it still seems appropriate to ask whether they are all as deluded and "irrational" as Flanagan and other scientific naturalists suspect.[4] The highly respected sociologist Peter Berger has argued that our everyday (prototypical) gestures of laughing, playing, and hoping—the very activities that keep us sane—could never occur if we were completely convinced deep down that death is the final word.[5] Once again, this is no proof that hope is realistic. In fact, naturalists today often explain hope away as a deceptive sur-

---

2. See Alfred North Whitehead, *Science and the Modern World* (New York: Free Press, 1925), 17. "Matter" here is really a name for the lifeless quantitative abstractions (especially what are called "primary qualities") of modern science.

3. William James, *Pragmatism* (Cleveland: Meridian Books, 1964), 76.

4. Sam Harris (*The End of Faith: Religion, Terror, and the Future of Reason* [New York: W. W. Norton, 2004]) is even more passionate than Flanagan in associating belief in life after death with irrationality.

5. Peter L. Berger, *A Rumor of Angels: Modern Society and the Rediscovery of the Supernatural* (Garden City, N.Y.: Doubleday, 1969).

vival mechanism, a Darwinian adaptation out of touch with reality. But scientific naturalists very seldom think through to the end—in the way that James does in the above quotation—what would be the full consequences of their confident claim that *absolute* death and nothingness await everything real.

Very few scientific materialists can embrace with full consistency the sober logic of the eminent physicist Steven Weinberg, who says that if there is no God and no life beyond death the most we can rescue from our absurd situation is a sense of honor in accepting this fate without flinching.[6] Yet, in the end, even Weinberg finds his life meaningful, for there can be no doubt that he believes implicitly that truth is worth seeking. Even the most pessimistic naturalists value truth, and their devotion to it energizes their lives. There seems to be great inconsistency here, for it is necessary to ask how thoroughly one can *value* anything that is taken to be destined for nothingness in the end. Furthermore, without at least a vague sense that our actions are intertwined with the eternal, as Teilhard declares, we could hardly sustain for long any real "zest for living."[7] It is not enough to settle for tragic heroism or a sense of honor in undertaking an aimless struggle, as sober naturalists such as Weinberg prescribe. Nor is it enough to understand action simply as a way of purifying our intentions in order to be right with God, as religious moralists have often taught. Instead what is needed is a sense that our actions, however insignificant, have an indelible impact on the universe and that something about this same universe lasts forever.[8]

As Teilhard has written, only a "passion *for being finally and permanently more*" can lead to a substantively ethical life, and this passion dies out if our efforts are felt as making no real difference in the end.[9]

> Man, the more he is man, can give himself only to what he loves; and ultimately he loves only what is indestructible. Multiply to your heart's content the extent and duration of progress. Promise the earth a hundred million more years of continued growth. If, at the end of that period, it is evident that the whole of consciousness must revert to zero, *without its secret essence being garnered anywhere at all*, then, I insist, we shall lay down our arms—and mankind will be on strike. The prospect of a *total death*

---

6. Steven Weinberg, *Dreams of a Final Theory* (New York: Pantheon, 1992), 255-56, 260.

7. Pierre Teilhard de Chardin, *Activation of Energy*, trans. René Hague (New York: Harcourt Brace Jovanovich, 1970), 229-44.

8. Ibid.

9. Pierre Teilhard de Chardin, *How I Believe*, trans. René Hague (New York: Harper & Row, 1969), 42.

(and that is a word to which we should devote much thought if we are to gauge its destructive effect on our souls) will, I warn you, when it has become part of our consciousness, immediately dry up in us the springs from which our efforts are drawn.[10]

Whitehead, no less ardently than James and Teilhard, ponders what it might mean if the universe as a whole were unable to attain some kind of immortality. Like Teilhard, he thinks that a consistent expectation of absolute death for the universe would trivialize, and eventually paralyze, human ethical aspiration. A persistent stimulus to the forming of Whitehead's own religious cosmology was his personal revulsion at the proposal by modern materialists that everything eventually comes to nought. The most coherent philosophy, he believed, is one that makes explicit room for the reality of something everlasting, something that might rescue *all* cosmic events, not only individual human lives, from ever perishing absolutely. This is why we shall always have to take religions seriously, in spite of their obvious defects. For at least religions give expression to our need for the eternal:

> Religion is the vision of something which stands beyond, behind, and within, the passing flux of immediate things; something which is real, and yet waiting to be realized; something which is a remote possibility, and yet the greatest of present facts; something that gives meaning to all that passes, and yet eludes apprehension; something whose possession is the final good, and yet is beyond all reach; something which is the ultimate ideal, and the hopeless quest.[11]

In stubborn opposition to scientific materialism, Whitehead argues that nothing actual can ever lapse completely into absolute oblivion. What is *really* going on in the universe is that everything worthwhile that happens, even though it may perish individually, is received into God's own unfading experience and endowed with a permanent significance. Our own efforts, too, no matter how futile they may seem at times, are not wasted. The cosmic past, though perhaps lost to our own memory, is forever preserved in the eternal immediacy of the divine experience, so nothing that ever happens can be lost absolutely. Everything has at least an *objective* immortality as it is received into the compassionate feelings of God (what Whitehead calls the "Consequent Nature of God"). Even though Whitehead is unsuccessful in stating how subjective immortality might be possible (a topic that I shall look into below), he

---

10. Ibid., 43-44.
11. Whitehead, *Science and the Modern World*, 191-92.

is able to address, at least partially, the materialist claim that the universe will dissolve into absolute nothingness.

In order to understand how such a proposal can avoid being labeled irrational, one must take seriously the recent revolutions in cosmology, physics, and biology that have shown the universe to be a process rather than a static mass of irreducible material particles. A processive universe, by definition, is made up not of material bits but of transient events, and these events or happenings can be related to each other in a temporal rather than simply spatial way. Every occurrence persists enduringly as an ingredient in those that follow. This means that everything that has happened in the past still has some impact on the present. The series of perished occasions still "matters" to all events that follow. It is in this temporally ordered, rather than a spatially frozen, way that the universe hangs together. The past, though faded and fixed, still adheres to the present and will also be part of the future. Passing events accumulate rather than vanish completely. In a processive universe things keep adding up, as it were, so significant events that happened in the remote past, such as the Big Bang or the origin of life, can still be "felt" at least faintly in present experience. Each present moment in any process is a "subject" that synthesizes the series of preceding events into itself in one way or other. In doing so it rescues the past from absolute perishing and makes it available to what is yet to come.[12]

Think of God, therefore, as the supreme subject that "stands beyond, behind, and within the passing flux of immediate things." God is real, but at the same time "waiting to be realized." That is, in a most vulnerable way, God is always being moved and even "changed" by what happens in the created world. This is not a denial of God's immutability, at least in any theologically relevant sense, for God's love and fidelity still abide forever, unshaken. It is precisely God's immovable love, expressed for Christians in the doctrine of the Trinity, that paradoxically allows God to be moved by what happens in the world.[13] God feels and thus saves the world by giving "meaning to all that passes," even while this divine graciousness "eludes apprehension."

---

12. This is a very brief synopsis and simplification of the philosophical cosmology presented in Alfred North Whitehead, *Process and Reality,* corrected ed., ed. David Ray Griffin and Donald W. Sherburne (New York: Free Press, 1968). For a useful introduction to the "process theology" based on Whiteheadian ideas, see John B. Cobb, Jr., and David Griffin, *Process Theology: An Introductory Exposition* (Philadelphia: Westminster, 1976). For a discussion of the ecological implications of process theology, see Charles Birch and John B. Cobb, Jr., *The Liberation of Life: From the Cell to the Community* (Cambridge: Cambridge University Press, 1988).

13. See Schubert Ogden, *The Reality of God and Other Essays* (San Francisco: Harper & Row, 1977), 47.

CAN GOD CHANGE?

The eternal God's vulnerability to what happens in the world, including both creative and tragic episodes, is consistent with our revelatory image of the divine descent. God becomes small in order to establish the most intimate relation to the world and to feel it and be moved by it in its tiniest detail. This is the responsive God of biblical faith. Moreover, the theme of divine futurity is also implicit in Whitehead's depiction of religion, inasmuch as God is "something whose possession is the final good, and yet is beyond all reach; something which is the ultimate ideal, and the hopeless quest." God, from the depths of an endless resourcefulness, continually offers the becoming universe relevant new possibilities for actualizing itself anew in each moment. God faithfully opens up the future by greeting the universe and the perishable lives within it with a wealth of possibilities for becoming new. Divine power consists, in part at least, of providing the possibility (*potentia*) for new being. Thus, God, in self-effacing humility and fidelity, acts powerfully and effectively in the world without violating the laws of nature. Such is the God, we may add, who can also raise the dead to new life.

God is not only the inspiration to novelty but also the *savior* of all the events that make up the cosmic drama. What happens in the world matters eternally to the self-giving love that we call God. In this sense God can be changed by what happens in the world. The Christian teachings about the dynamism of the Trinity are a way of expressing this intuition. As the Father, God in generative love first arouses the universe to new being; as the Son, God gives the divine fullness to the world irreversibly and forever; and then, in the Spirit, God assimilates into the divine life all the transience in the cosmos that would otherwise mean eternal loss. Incarnation, redemption, and eschatological hope, in their cosmic ramifications, also imply that God is affected forever by events in the physical world.[14]

Theology in an age of science cannot do justice to faith's claims about God's kenotic love unless it allows that the divine Mystery can undergo change and even suffering precisely because God remains immutably faithful to the covenantal promise. Teilhard, though less explicitly than Whitehead, also allows that God is truly altered by what happens in the world, and in such a way as not to diminish the divine perfection in any way. God, he says, is in a sense self-sufficient, "and yet the universe contributes something that is

---

14. As I will argue below, this compassionate embrace of the universe must also include the saving and transforming of our subjective consciousness beyond death.

vitally necessary to him."[15] How could it be otherwise if God truly loves the world enough to take on its very materiality in the incarnation. What happens in the world must matter forever to a God of limitless love. "My wanderings you have counted; my tears are stored in your flask; are they not recorded in your book?" (Ps 56:9). So we may trust that even our smallest efforts and experiences can have cosmic significance since along with all the world's occurrences, including the entire evolution of life, they are taken everlastingly into the immediacy of God's experience. A resurrection faith allows us to hope that even tragic and evil occurrences in natural and human history will never be forgotten but can instead be given a redemptive meaning by assimilation to the cross of Christ and the trinitarian drama of salvation.

FINDING MEANING IN AN UNFINISHED UNIVERSE

The fact that creation is still going on and the universe remains unfinished bestows on us, and on other creatures too in their own way, the dignity of being co-creators with God. A counterfactual alternative to the unfinished universe that science has revealed might be one in which God is the sole creator and no room is left for significant human or creaturely effort. But such a quietistic perspective, one that prescientific philosophies and theologies often espoused, provides little basis for the zest for living that Teilhard is idealizing. Human life, as he rightly insists, can best be energized by a sense that there is still room for *more being*. Otherwise hope is suffocated and human action rendered pointless from the start.

If the universe were thought of as essentially finished or perfect, what purpose would our efforts have except perhaps the purifying of intentions? This may have seemed good enough for many religious believers in the past, but the modern secularistic rejection of Christianity should by now have instructed theology that classic spiritualities advocating indifference to the earth's future and the universe's own destiny no longer evoke excitement in most people. Religious worldviews that promote or tolerate a sense of the cosmic futility of human efforts fail to motivate. Human beings need hope in order to live lives of passion and ethical vigor. But this means that they also need a universe that still has room to grow, to become more. In other words, they need an unfinished universe, and this is exactly what science has given us.

---

15. Pierre Teilhard de Chardin, *Christianity and Evolution*, trans. René Hague (New York: Harcourt Brace Jovanovich, 1969), 177.

An initially perfected universe could have no future, no room for "more being," and this would only "clip the wings of hope."[16] For this reason contemporary theology would do well to take as its starting point the unspeakably good news of God's liberating humility and futurity that give rise to a cosmos still coming into being.

According to the theology of nature I am setting forth here, God selflessly refrains from rounding off the universe *in principio,* since to have finished it at once would have left the world without a future. God lovingly "withdraws" from overwhelming each present moment with the divine infinity, instead taking up domicile as the world's all-replenishing future. This understanding of God as "essentially future" entails in no sense a deistic dismissal of the divine from present involvement in the world. In fact just the opposite is the case. It is by arriving from the future that God brings new being to each present moment, rescuing it from being swallowed up by the dead past. A deeper form of involvement is hard to imagine.

To carry this thought a bit further, as I noted above, everything that has already happened in the universe still remains immortally ingredient in each present event, however faintly, so the past too becomes subject to eternal renewal in the context of the divine futurity. This theological scheme, I believe, allows for a religiously robust sense of divine care—a doctrine of providence—without having to contradict scientific discovery. God is deeply and constantly involved in the cosmic process, as both its inspiration and its savior, but in such a way as not to tinker with natural laws or make constant adjustments to the physical world. God, the world's future, remains faithfully available to the cosmos as the infinite reservoir of new possibilities upon which evolution and human creativity can draw in order to give definition to the world.[17] God's self-emptying restraint and generous futurity are so intimately interwoven with the processive universe that they go largely unnoticed except to the eyes of faith.

The entire universe (or multiverse), as it sweeps narratively across vast epochs of time, is continually received into the compassionate embrace of the everlasting Trinity. We may think of God's Spirit as the ultimate power of renewal that continually places the world in a "free and open space" with an ever-new future up ahead. "To experience what is *ruach* [Spirit]," Jürgen Moltmann writes, "is to experience what is divine not only as a person, and

---

16. Ibid., 79.

17. If theology wants to make sense of miracles in a scientific age it may think of them not as magically breaking the laws of nature, but as events, fully consistent with natural laws, that open up the future in a decisive way.

not merely as a force, but also as *space*—as the space of freedom in which the living being can unfold."[18] "God the Father" refers here to the infinite generativity out of which new possibilities are always made available to the universe so that it may undergo renewal by the power of the Spirit. And God the Son, through the incarnation, concretizes the divine descent in nature and history, gathering all things corporeally to himself and handing them over to the Father, again by the power of the Spirit. There is the basis here also for a eucharistic theology in which our consumption of the body and blood of Christ signifies our being assimilated into the universe which by extension is also the body of the Savior. "Like a powerful organism," Teilhard writes, "the world transforms me into him who animates it. 'The bread of the Eucharist,' says St. Gregory of Nyssa, 'is stronger than our flesh; that is why it is the bread that assimilates us, and not we the bread, when we receive it.'"[19]

Even though the whole series of cosmic events is subject to eventual temporal perishing, its constituent moments are constantly streaming into the refinery of creative transformation by the Spirit of God. In our theological understanding of this processive world there is abundant death and perpetual perishing, but there is also redemption, preservation, and new creation. Hence, what Christians refer to as resurrection, an occurrence that we believe to be bodily and physical, is in no sense incompatible with contemporary science, even though it is incompatible with scientific naturalism. We may understand human bodily resurrection, minimally at least, as the reception into God's life of each person's story.[20] Whatever else resurrection might mean, at the very least it entails the solidarity of our own personal stories with that of Jesus, and this communion in turn is received into the eternal divine drama that we refer to as Trinity. But persons, as we now realize, are inseparable from a universe composed of events that occur in the present and are then thrust into the past as new possibilities keep arriving from out of the future. As its constituent events keep adding up, the whole cosmic story is received everlastingly into the vivifying immediacy of God's experience. And since everything that happens in our personal lives is woven into the fabric of the whole universe in an unrepeatable way, each person's life is taken along with the whole cosmic story into God.

---

18. Jürgen Moltmann, *The Spirit of Life: A Universal Affirmation,* trans. Margaret Kohl (Minneapolis: Fortress, 1992), 43.

19. Pierre Teilhard de Chardin, "My Universe," in *Science and Christ,* trans. René Hague (New York: Harper & Row, 1965), 75-76; see also idem, "The Mass on the World," in Thomas M. King, *Teilhard's Mass: Approaches to "The Mass on the World"* (New York: Paulist, 2005), 145-58.

20. See Hans Küng, *Eternal Life: Life after Death as a Medical, Philosophical and Theological Problem,* trans. Edward Quinn (Garden City, N.Y.: Doubleday, 1984), 110-12.

But what if the universe, considered as a whole, will perish, as contemporary cosmology and the laws of thermodynamics predict? This is an especially serious issue since, as Whitehead notes, the ultimate evil in the world is the simple fact that things perish.[21] That the entire universe will eventually be lost is a most sorrowful prospect. However, there is no reason for theology to be any more surprised that the universe will perish than that any particular thing in it will eventually perish. Indeed, Christian theologians should already have realized that everything other than God is perishable. They should not be too taken aback by current astrophysical predictions of a cold collapse of the originally hot Big Bang universe trillions of years from now. As long as the "secret essence" of the universe and consciousness is being "garnered" somewhere everlastingly, as Teilhard proposes, the cosmos need not be thought of as ultimately "pointless," even if it will collapse eventually into an energetic deep freeze. The everlasting care of God can surely save the perishing world as a whole, in the manner described above, recording it and reordering it continually into wider patterns of beauty in the vision of divine glory that we "hope to enjoy forever."

THE QUESTION OF SUBJECTIVE IMMORTALITY

What about our own subjective survival of death? Sad as it may be at first, Flanagan replies, scientifically enlightened rationality now obliges us to abandon any such expectation. Science has shown, he maintains, that hope in survival beyond death is completely illusory. Along with other evolutionary naturalists, Flanagan claims that Darwinian insights have decisively exposed the fanciful character of all human longing for conscious immortality.[22]

According to Darwinian naturalists, the *real* reason our denial of death's finality is so deeply ingrained in our nature—as a "prototypical gesture," in Peter Berger's terminology—has nothing at all to do with the actual existence of a transcendent dimension where the soul's survival could be sealed forever. There is a much simpler, purely natural explanation for such naive trust. The *ultimate* reason for the allure of ideas of resurrection and immortality is that they are adaptive in an evolutionary sense.[23] Hope for subjective immortality,

21. Whitehead, *Process and Reality*, 340.

22. Flanagan, *Problem of the Soul*, 301-10; see also Daniel C. Dennett, *Darwin's Dangerous Idea: Evolution and the Meaning of Life* (New York: Simon & Schuster, 1995).

23. Robert Hinde, *Why Gods Persist: A Scientific Approach to Religions* (New York: Routledge, 1999); Walter Burkert, *Creation of the Sacred: Tracks of Biology in Early Religions* (Cambridge, Mass.:

according to evolutionary anthropology, has led people to *believe* that they have an eternal worth. This conviction in turn has given them a reason to live well, to marry and have children, and thus to allow human genes to attain a kind of immortality. Indeed, long ago—probably during the Pleistocene period one to two million years ago—hominid genes began to engineer brains that could entertain thoughts about subjective immortality, and these thoughts eventually became entrenched in most religions, including Christianity. It is our species' inherited genomic configuration that inclines people even today to hope for conscious existence after they die. According to evolutionary naturalists this hope can now be explained quite economically in terms of the Darwinian idea of adaptive functioning.[24]

Religious ideas such as immortality persist into the present, evolutionary psychologists claim, because they are adaptive. The propensity to extend ourselves imaginatively toward endless life beyond the grave lies buried deeply and perhaps even permanently in the digital genetic instructions that shape our nervous systems. Hope for eternal life is a kind of trickery that strands of DNA enact in order to get themselves passed on to future generations. Consequently, since it now appears that the roots of religion lie more foundationally in biology than in culture, many Darwinian naturalists no longer expect to eradicate completely our pious dreaming about immortal existence. Even in an age of scientific enlightenment phantasms of subjective survival will stubbornly linger on. Some evolutionists are even quite forbearing toward our naive longing for subjective immortality even while they dismiss it as childish fiction that we need to outgrow.

There is an undisguised element of condescension here, of course, but this recent brand of evolutionist explanation is not as uncompromisingly dismissive of religious hope as were earlier versions of scientific naturalism. Darwinian debunkers now acknowledge that we all have the same genetic makeup as our religiously deluded ancestors. And sometimes they even seem grateful that religious illusions have existed until now, since indirectly these fantasies have helped keep human genes around long enough to make our own existence possible.[25] Nevertheless, Darwinian naturalists do not want us

---

Harvard University Press, 1996); Pascal Boyer, *Religion Explained: The Evolutionary Origins of Religious Thought* (New York: Basic Books, 2001).

24. For a sustained critique of such rationalization, see Holmes Rolston III, *Genes, Genesis and God: Values and Their Origins in Natural and Human History* (New York: Cambridge University Press, 1999).

25. See, e.g., Loyal Rue, *By the Grace of Guile: The Role of Deception in Natural History and Human Affairs* (New York: Oxford University Press, 1994), 82-127, 261-306. At times Rue even refers to religious ideas as "lies," but he sees little harm in people still believing in them.

to forget that faith in subjective survival beyond death is still epistemologically deluded, even if it is biologically productive.

THEOLOGY AND SKEPTICISM IN AN UNFINISHED UNIVERSE

Even before Darwin's own ideas had taken root, however, the English poet
Alfred Lord Tennyson's moving poem "In Memoriam" had already expressed
his age's great uncertainty about the implications of science for hope in resurrection:[26]

> Behold we know not anything;
> I can but trust that good shall fall
> At last—far off—at last, to all,
> And every winter change to spring.
>
> So runs my dream: but what am I?
> An infant crying in the night:
> An infant crying for the light:
> And with no language but a cry.
>
> O life as futile, then, as frail!
> O for thy voice to soothe and bless!
> What hope of answer, or redress?
> Behind the veil, behind the veil.[27]

A theology of nature must be especially sensitive to the anguish expressed
in this desperate longing for an eternal light now hidden "behind the veil."
Most people can identify with Tennyson's blend of hope, doubt, and loss. And
most of us would agree that Whitehead's sense of a purely objective kind of
immortality is simply not enough to soothe our anxiety in the face of death.
We want to know whether science definitely rules out *subjective* survival, as
Flanagan and other naturalists claim to be the case.

But why does our own final personal destiny appear so uncertain in the
first place? I would suggest here that an often overlooked reason for our
wavering on this matter is that the universe, out of which our lives and our

---

26. William E. Phipps notes that even before Darwin's *Origin* was published, Tennyson had
already become familiar at least with the controversial evolutionary ideas of Robert Chambers, *Darwin's Religious Odyssey* (Harrisburg, Pa.: Trinity Press International), 95.

27. Alfred Lord Tennyson, "In Memoriam." For a readable study of nineteenth-century British
skepticism, see A. N. Wilson, *God's Funeral* (New York: W. W. Norton, 1999).

religious aspirations have sprouted, is itself still in the process of coming into being. If the cosmos were completely finished, we might legitimately expect the veil to fall and clarity to shine through, as Tennyson requests. But the uncertainty within all religious hope is itself, at least in some way, a correlate of the fact that humans, with all their aspirations, are part of an immense cosmos still in the making.[28] We need to keep this cosmic setting in mind as we speculate in a scientific age on the prospect of conscious life after death. Indeed, the existence of scientific skepticism itself may be seen as symptomatic of the unfinished state of things. Since creation itself is not yet fully actualized, and hence not yet fully intelligible, it is not surprising that we can discern both God and our destiny only dimly. Teilhard even proposes that the darkness in which faith moves and the incompleteness of the cosmos itself are inseparable from the general problem of evil.[29] Only in the final vanquishing of suffering and death can we expect to see clearly. Now is the time, therefore, to decide between hope and despair since clarity is not an option for anyone.

The scientific materialist, however, wants no ambiguity at all here and now. Flanagan, for example, declares with all the self-assurance of any true believer, that "when we die, we—or better, the particles that once composed us—return to nature's bosom, not to God's right hand."[30] He goes on to proclaim that beyond what naturalism is able to discern there lies absolutely nothing:

> If you wish that your life had prospects for transcendent meaning, for more than the personal satisfaction and contentment you can achieve while you are alive, and more than what you will have contributed to the well-being of the world after you die, then you are still in the grip of illusions. Trust me, you can't get more. But what you can get, if you live well, is enough. Don't be greedy. Enough is enough."[31]

Perhaps if our universe were now complete and finished, one could rightly demand such an abrupt end to all uncertainty. But, as long as the world is still coming into being, and we and our minds remain fully tethered to an unperfected cosmos, can we expect to see our way as clearly out of the mist as Flanagan's naturalism desires? I prefer to go along with Whitehead when he says that "the greatest of present facts" lies beyond all reach. In making the world "safe for naturalism," Flanagan, however, wants to eliminate all haziness. He would undoubtedly approve the famous lines of Algernon Charles

---

28. This is a point made especially well by Teilhard in *Christianity and Evolution*, 81-84.
29. Ibid.
30. Flanagan, *Problem of the Soul*, ix-x.
31. Ibid., 319.

Swinburne, who lived in the same uncertain century that gave us Tennyson. Shaking off all doubt, this poet announces not only that our complete perishing is certain but that we should be grateful for it:

> We thank with brief thanksgiving
> Whatever Gods may be
> That no life lives forever;
> That dead men rise up never;
> That even the weariest river
> Winds somewhere safe to sea.[32]

During the last century and a half, more than a few enlightened minds have moved from Tennyson's anguished hesitancy to Swinburne's and Flanagan's crystalline certitude about the finality of the grave. Still, most humans continue to trust, even if only tacitly, that death is not a definitive ending to conscious awareness. For many, the prospect of an absolute dissolution of consciousness would entail the final victory of evil over good. In any case, in the light of current cosmology, unless we wish to remain committed dualists, the question of our personal survival of death cannot be divorced from that of the universe's destiny and the larger question of evil and redemption. Unlike cosmic pessimism, the posture of Christian hope is one in which the entire universe has a hidden meaning that—because the world is still in the process of being created—is at least partially hidden from view here and now.

THE REALISM OF HOPE

It is hope, rather than cosmic pessimism, naturalistic certitude, or dualistic withdrawal that matches up most naturally, and I believe realistically, with the unfinished universe in which we now find ourselves. Evolutionary biology and other natural sciences have demonstrated beyond any reasonable doubt that we humans are indeed fully part of an evolving universe. So theology's traditional preoccupation with individual eschatology can be insulated only artificially from a deeper and wider concern about the cosmos as a whole and where it might end up. It is one of the felicitous consequences of today's science that in its presence theology can no longer plausibly separate the issue of personal destiny from the larger topic of the universe's final outcome. However, whether contemporary Christian eschatology will begin to pay more attention to this linkage remains to be seen.[33]

---

32. Algernon Charles Swinburne, "The Garden of Persephone."
33. A notable exception to such neglect is Anthony Kelly, C.Ss.R, *Eschatology and Hope*, Theology in Global Perspective (Maryknoll, N.Y.: Orbis Books, 2006).

The doctrine of immortality of the soul, we need to recall, has flourished primarily in the context of an otherworldly, acosmic spirituality. Historically, belief in human survival beyond death seems to have fit most comfortably into a metaphysics that views the transient world of nature as itself essentially pointless apart from its serving as the stage and backdrop for the human drama of salvation. Accordingly, our final resting place has often been pictured as a timeless spiritual realm situated far apart from the physical universe. Even today countless devout people assume that the only alternative to fatalistic materialism is an anticipation of the soul's final withdrawal from any contact with the cosmos at all. The roots of this otherworldly optimism lie in Christianity's early experiments with Platonic thought. Jürgen Moltmann summarizes this ambiguous heritage:

> In the degree to which Christianity cut itself off from its Hebrew roots and acquired Hellenistic and Roman form, it lost its eschatological hope and surrendered its apocalyptic alternative to "this world" of violence and death. It merged into late antiquity's gnostic religion of redemption. From Justin onwards, most Fathers revered Plato as a "Christian before Christ" and extolled his feeling for the divine transcendence and for the values of the spiritual world. God's eternity now took the place of God's future, heaven replaced the coming kingdom, the spirit that redeems the soul from the body supplanted the Spirit as "the well of life," the immortality of the soul displaced the resurrection of the body, and the yearning for another world became a substitute for changing this one.[34]

But there is another possibility. It is that of a patient, long-suffering hope for the *whole* universe, a universe that includes persons with aspirations of existing consciously and everlastingly in the presence of God's unfathomable beauty. Over the past half century, science has shown without a doubt that the emergence of the religious species of mammals is continuous with other stages in natural history. So also is the incarnation of God in human history. "The Flesh of Christ," as Teilhard nicely puts it, "is fed by the whole universe. The mystical Milieu gathers up everything that is made of energy."[35] Furthermore, as seen in the context of nature's larger history of emergence, the present cosmos may still be young enough that there are other creative surprises awaiting in the future. There may well be a lot more that still has to happen in this world before our souls should begin to think about migrating elsewhere.

---

34. Moltmann, *Spirit of Life*, 89.
35. Teilhard de Chardin, *Science and Christ*, 77.

Recently astrophysics has extended the chronicle of the forging of consciousness back to at least the moment of the Big Bang. And since the universe may have an even more unfathomable depth of future time ahead of it, there is no clear reason to assume that consciousness itself will not undergo significant transformations in the cosmic future.

Is it not then a bit premature for us humans to be obsessed exclusively with the question of personal subjective immortality? Instead should not theology be concerned primarily with the survival of the larger cosmic adventure of consciousness? What if Teilhard is right when he "sees" a planetary consciousness, a noosphere, now just beginning to form here on earth? Perhaps, too, there are other similar noospheres forming extraterrestrially.[36] In any case, contemporary cognitive science, astronomy, astrobiology, and cosmology invite us to widen our eschatological hopes for conscious survival far beyond those of our classic earthbound theology textbooks.

The question, then, is what will happen ultimately to consciousness on a cosmic scale, not just to the very limited cognitional centers that we call intelligent subjects here on our tiny planet. And where will the totality of consciousness emergent in cosmic history, not just our own, have gone when our universe itself has faded away? Will it all have simply disappeared into the void? Admittedly, in the face of the universe's own eventual "death" it is hard not to remain entranced by individualistic supernaturalist optimism. Soul-centered eschatologies of traditional Christian theology allow us to surmount cleanly the universe's perishability. However, alluring as classic notions of the harvesting of souls from a "vale of tears" may seem in our moments of loss, this brand of otherworldly optimism still seems unrealistically aloof from new scientific knowledge of the universe and our intimate confluence with it. Pre-scientific notions of immortality may fail to measure up to the vision that God may have for the fulfillment of the larger epic of consciousness. Since here and now we have no existence, even as conscious subjects, apart from the fabric of the universe, eschatology must ask whether subjectivity can ever be separated fully from a cosmic setting. Human existence here and now is irreducibly communal and cosmic. It would seem strange if this were not the case

---

36. See Teilhard de Chardin, *Activation of Energy*, 99-127. In 1944 Teilhard wrote that there is a "positive likelihood," of other planets inhabited by intelligent beings. In that case "the phenomenon of life and more particularly the phenomenon of man lose something of their disturbing loneliness." For all we know there may be many "noospheres" or "thinking planets." "It is almost more than our minds can dare to face," he says, but the evolutionary tendency toward complexification, consciousness, and centration may be "cosmic" in scope. Nevertheless, "there can still be only a single Omega," that is, a single transcendent God whose being embraces and awaits the entire evolving universe" (*Activation of Energy*, 127).

eschatologically also. The question of conscious personal survival remains inseparable from interest in the ultimate outcome of the cosmic story.

In summary, if the universe as a whole were a futile drift toward absolute extinction, the significance of our own personal lives would be placed in question also. And if human subjectivity is inseparable from the natural world, we simply cannot dualistically divorce our own personal search for significance and survival from the question of what is going to happen to the entire universe. Perhaps we could have done so at a time when it seemed that our own existence and consciousness were only tenuously and accidentally connected to the physical world. But evolutionary biology, geology, neuroscience, and astrophysics no longer allow us to think realistically even of human subjectivity as not belonging fully to the universe.

> All around us [says Teilhard], until it is lost to sight, radiates the net of spatial and temporal series, endless and untearable, so closely woven in one piece that there is not one single knot in it that does not depend upon the whole fabric. God did not will individually . . . the sun, the earth, plants or Man. He willed his Christ;—and in order to have his Christ, he had to create the spiritual world, and man in particular, upon which Christ might germinate;—and to have man he had to launch the vast process of organic life . . . ;—and the birth of that organic life called for the entire cosmic turbulence.[37]

When Teilhard wrote these lines in 1924, he had no idea of how decisively Big Bang cosmology, quantum physics, astrophysics, and biology would bear out his intuitions.

BUT WILL "I" SURVIVE?

Once again, what can we say about conscious subjective human existence beyond death in an age of science? Flanagan's ideal of temporary terrestrial "flourishing" as being "enough" will have little appeal to those whose lives are not as comfortable and secure as that of the average twentieth-century American academic. And an exclusively objective immortality would, for most people, differ only insignificantly from Flanagan's naturalism. Any Christian theology faithful to the tradition and in touch with biblical hope must make conceptual room for subjective immortality along *with* cosmic destiny, not separate from it. To do so, it is well advised to start with the idea of a God

---

37. Teilhard, *Science and Christ*, 79.

who is *absolutely related* to the universe and in whom the totality of cosmic events is internalized everlastingly.[38] But it must go beyond this. A theological defense of the reasonableness of hope for subjective immortality must base itself primarily not on ancient anthropology or contemporary cosmology but on the trustworthiness of God.

Theology needs to ask whether God could command our complete trust if at the same time the Creator were thought of as presiding over the complete extinguishing of the conscious kind of subjectivity that allows humans to live in relation to God in the midst of the universe's perpetual perishing. Could we sincerely maintain that we will truly be saved everlastingly if our consciousness is to lapse into eternal sleep? At least as far as the individual is concerned, a purely objective immortality would not differ much from the eternal silence that William James refers to when considering the implications of scientific materialism, namely, that "nothing, absolutely nothing remains." To elicit our worship and confidence in the face of all loss, the divine power that awakens and sustains conscious human subjectivity within the limits of a perishable, unfinished world must be able to do so also in a consummated one.

ESCHATOLOGICAL PANVITALISM

A major question for theology in the age of science, then, is whether there is a reasonable alternative to the intellectual appeal of the ontology of death out of which Flanagan's naturalistic assumptions emerge. And could such an alternative permit faith's openness to resurrection without contradicting scientific understanding and knowledge? What I propose is an *eschatological panvitalism* based on the impression that nature is not simply and solely the outcome of a past series of mechanical causes, but also the *anticipation* and *promise* of an indeterminate cosmic future—including eventually a decisive and final victory of life over death, and consciousness over unconsciousness. This speculation is inspired by religious hope, of course, but it is not inconsistent with certain stirrings in the world of recent scientific inquiry wherein nature is observed to be an emergent process whose various states suggest the attraction of something up ahead rather than a compulsion arising completely from the past.[39]

---

38. A doctrine made most explicit in Charles Hartshorne's book *The Divine Relativity* (New Haven: Yale University Press, 1948).

39. See, e.g., Harold J. Morowitz, *The Emergence of Everything: How the World Became Complex* (New York: Oxford University Press, 2002).

Eschatological panvitalism entails a metaphysics based on the sense of nature's promise of indeterminate outcomes yet to come, rather than exclusively on a sense of what has already occurred. Contrary to the modern ontology of death, moreover, the worldview I am proposing is one that takes aliveness to be the *essential* characteristic of nature—and consciousness an essential characteristic of being—even if such a condition is not obvious or fully actual yet. The eschatological character of Christian faith encourages theology to declare that the essential state of nature has yet to be realized and made fully visible. This means that life can be understood as irreducible to death if our thoughts are directed *forward* to where cosmic process may be going, rather than exclusively backward to where it came from. Such a reversal of vision, I believe, is one of the great contributions of biblical and Christian faith to an understanding of the world.[40] In saying this, I am also calling attention once again to the importance of Teilhard's point that the world hangs together not by mechanical power transmitted from the past but by an attracting future up ahead, a realm of yet unrealized possibilities wherein coherence, and hence intelligibility, can finally become complete. At the origins of Christian tradition only those who were open to God's coming, especially those whose past was too inglorious to be the basis of their self-esteem, had their sight transformed by hope in such a way as to be adequate to the news of their Lord's resurrection. In a parallel way we today can really know life only by looking forward toward its full future actualization rather than by gazing backwards toward its remote physical past or its lifeless elemental components.

The final word about the truly *essential* state of nature cannot be arrived at simply by tracing life's antecedents into the dead cosmic past. A metaphysics based exclusively on a survey of the cosmic past, after all, is likely to yield only an ontology of lifelessness. But by acknowledging the *anticipatory* character of nature perhaps we can finally begin to get a bead on life. For Christian faith, Jesus' resurrection is the revelation of what nature anticipates, a fulfillment in which life will show itself at last to be more fundamental and ultimately more intelligible than death. This is why our trust in the news of resurrection gains little support from our looking to the past for its cognitive foundations. Contrary to the beliefs of scientific materialists, it is not death but life that will in the end prove to be the more intelligible state of being, since the imagined deadness of the material past dissolves into a formless incoherence the deeper we delve backward into it. Resurrection, therefore, is not an unintelligible

---

40. See Pierre Teilhard de Chardin, *The Human Phenomenon*, trans. Sarah Appleton-Weber (1959; Portland, Ore.: Sussex Academic Press, 1999), 1-3.

interruption of nature but the final vanquishing of deadness and disunity. Thus, it is by anticipating nature's essential, though not yet actualized, eschatological aliveness, and not by peering backwards and downwards toward the earlier-and-simpler antecedents and physical constituents of life, that theology will be able to arrive at an accurate reading of the cosmos. Such an approach to understanding the resurrection will have the additional advantage of not conflicting in any way with natural science.

A Christian theology of nature, moreover, must no longer compromise or tacitly cooperate with the modern ontology of death. Rather it must look more enthusiastically toward death's final overcoming. To those who believe that God is the Author of all life, as well as the One who raises the dead to new life, the intuitions of primal panvitalism must somehow be correct after all. Along with our ancestral panvitalists, theology may trust that reality is *essentially* alive and only provisionally lifeless. But in the face of the quantitatively overwhelming inertness of the cosmos in its temporal past and its present spatial outreach, panvitalism can best be taken as referring to the world's ultimate future. For those who put their trust in the God of biblical revelation, death cannot be the normal, natural, final, or most intelligible state of things. If we believe in a God whose enemy is death we cannot make peace with any ideology that cedes ontological primacy to what is lifeless.[41]

A faith shaped by the hope for resurrection, therefore, will be especially critical of the ideology of death that scientific naturalism usually presents as pure scientific fact. It must also resist the compromises with scientific naturalism that the arguments of proponents of intelligent design make in order to find a place for life in the midst of what they assume to be the prevalent deadness of nature. Given the intellectual dominance of scientific naturalism, intelligent design opponents of evolutionary biology rightly look for a way to reconnect a lifeless cosmos to a living God. Theologically speaking, however, intelligent design provides only unsteady refuge. Not only does it embarrassingly introduce theological categories into scientific accounts of life; what is worse, it only half-heartedly challenges the ontology of death that has become the worldview tacitly presupposed by contemporary scientific naturalism. It still allows for the final rule of lifelessness throughout what it takes to be the inanimate world.

---

41. It is true that the perishing of organisms is an evolutionary necessity, in order to make room for sufficient genetic diversity. Moreover, it is also true (as theologian Elizabeth Johnson has reminded me) that there is a sense in which the spiritual journey may include the befriending of death. Even Teilhard de Chardin advises us that we must embrace not only our activities but also our passivities (including suffering and eventually death). All I am emphasizing here is that theology cannot accept death as ultimate, either in the explanation of life or as the final state of being.

The dominant biblical worldview is one in which life has primacy. But such a metaphysics can be justified only by allowing that the fullness of life, intelligibility, being, and consciousness belongs to the future. Right now, here in the present, we are witnesses to the ambiguity and perishability of life as well as to what may seem to be the finality of death, especially if our eyes are focused either on the present or straining only toward nature's past. We cannot claim on scientifically empirical grounds alone that life is the norm and death the exception. If we are panvitalists it can only be of an eschatological variety. The end of all life, and death, must be life in abundance. But the fullness of life, at least here and now, is less fact than promise. If we confess continuity with our biblical past we must find a way to reaffirm here and now that life is the essential and most intelligible state of being. But from the perspective of the present, the essential has not yet become fully actual. In other words life, at least in some important sense, has not yet happened.

A resurrection faith, I believe, pushes us toward a metaphysics of the future. That is, it implies that what is most real or essential, when seen from our present perspective, can take hold of us only as we turn toward what is not-yet. And we can know the future not by grasping it the way we do things that lie apparently finished in the past, but only by allowing ourselves (and nature) to be carried away by it. By adopting the posture of hope, in other words, we can begin to approach the realness of life.

The foundational state of nature is not the dead past but the future on which it leans "as its sole support."[42] What is called for is a whole new metaphysical setting for science, one that will encourage further research but not ask us to throw away either common sense or religious intuition. The very same universe that some prominent scientific naturalists have characterized as aimless and pointless becomes rich with purpose as soon as we understand its Creative Source to be a self-emptying love that perpetually comes toward the present from out of the future.[43] I would suggest, then, that envisaging God as essentially future, and nature as promise (rather than as either perfectly designed or essentially absurd) is a fertile enough framework to embrace both science and a theology of resurrection. We no longer need to look at the emergent fact of life as mere veneer temporarily concealing an ultimate deadness. Rather, the present reality of life is the harbinger of an unimaginable future now being opened up by the Spirit of the God who raised Jesus from the dead.

---

42. Teilhard, *Activation of Energy*, 239.
43. See Karl Rahner, S.J., *Theological Investigations*, vol. 6, trans. Karl and Boniface Kruger (Baltimore: Helicon, 1969), 59-68.

SUGGESTIONS FOR FURTHER READING AND STUDY

Berger, Peter L. *A Rumor of Angels: Modern Society and the Rediscovery of the Supernatural.* Garden City, N.Y.: Doubleday, 1969.

Flanagan, Owen. *The Problem of the Soul: Two Visions of Mind and How to Reconcile Them.* New York: Basic Books, 2002.

Küng, Hans. *Eternal Life: Life after Death as a Medical, Philosophical and Theological Problem.* Translated by Edward Quinn. Garden City, N.Y.: Doubleday, 1984.

Moltmann, Jürgen. *The Coming of God: Christian Eschatology.* Translated by Margaret Kohl. Minneapolis: Fortress, 1996.

10

Scientific Truth and Christian Faith

A FUNDAMENTAL QUESTION IN THE DIALOGUE of science with religion is whether faith is realistic, and revelation true. Scientific naturalism, impressed as it is by inductive reasoning, rejects the claims of Christianity as unreliable since they are empirically untestable. The naturalistic assumption is that if any propositions are to be widely accepted as true they must conform to publicly available criteria of right knowing as understood by science and scientifically influenced philosophy. But since the ideas associated with Christian revelation are not congenial to public, objective confirmation, the naturalist is required to question their cognitive standing.

Up to this point I have been exploring the *meaning* of the new scientific pictures of nature as seen in the light of the revelatory image of God's descent and futurity. But to people who are impressed with science and its experimental method, the question of truth is often of much more concern than that of meaning. They want us to ask whether in allowing ourselves to be grasped by the power, goodness, and beauty of revelation, we can be confident also that we are being taken captive by *truth*. Is it conceivable perhaps that our trust in a God of promise and self-emptying love is a grand illusion? Is the revelatory image that I have been taking as foundational to a theology of nature one that opens the inquiring human intellect to an endless journey of discovery? Or is it instead a dead end that suffocates the mind and closes it off to the way things really are?

If there is truth in the claims of Christian revelation, then the life of faith should have the effect of opening one's mind to the totality of being, including the natural world. There must be a positive relationship between the effect that trust in God has on the believer's mind and the same mind's willingness to open itself to the results of scientific inquiry. However, scientific naturalists strongly deny that faith can support the quest for scientific truth, and they nearly always point to the ways in which Christians have closed their minds in modern times to new scientific discoveries, especially those of Galileo and Darwin.[1]

---

1. That Christian teaching has been the main obstacle to the advance of science historically is the

Today they may wonder why so few devout Christian believers are numbered among the scientifically elite. For example, how many self-avowed Christian believers can be found on the list of those who hold membership in the National Academy of Sciences (a high percentage of whom are self-described atheists)?[2]

One may reply that most of the great minds that laid the foundations of the modern scientific movement (Copernicus, Galileo, Descartes, Newton, and Boyle) were devout theists, but naturalists these days will not be impressed. Science arose in spite of their faith, they claim, not because of it. The life of faith clouds the mind so as to keep it in darkness about what is really going on in the world. Faith dulls, if it does not destroy, intellectual curiosity about the true nature of the physical universe and life. Above all, it causes people to reject the harsher implications of evolution and cosmology.[3] Whether justifiable or not, this is the impression Christian faith leaves on many scientifically enlightened critics today.

A theology of nature, therefore, must do more than look for the theological meaning of scientific discoveries. It must also demonstrate that trust in the content of revelation can actually *support* the mind in its quest for scientific truth. To do so would be an implementation of what I referred to earlier as the "confirmation" approach, and this is the task I shall undertake in the present chapter. Of course, making a case for the truthfulness of revelation will seem futile to scientific naturalists. They will invariably insist that, in order to be accepted as true, revelation must be independently verifiable by science. However, theology has good reason to insist that it is the very nature of revelation to reside beyond the scope of scientific certification. Inevitably scientific naturalists will answer back that this is an evasion, and they will go on to insist one more time that nothing truly real can exist beyond the potential reach of science. But a justifiable rejoinder is that this scientistic claim by naturalists cannot be scientifically confirmed either. It is no less a matter of belief

---

well-known and controversial thesis of Andrew Dickson White (*A History of the Warfare of Science with Theology in Christendom* [New York: Free Press, 1965]) and John William Draper (*History of the Conflict between Religion and Science* [New York: D. Appleton, 1898]).

2. In a recent survey of National Academy of Sciences members fewer than 10 percent of those who responded professed to believe in a personal God, and among biologists only 5 percent did so (E. J. Larson and L. Witham, "Scientists and Religion in America," *Scientific American* 281 [1999]: 88-93).

3. See, e.g., Carl Sagan, *The Demon-Haunted World: Science as a Candle in the Dark* (New York: Random House, 1995); Michael Shermer, *Why People Believe Weird Things: Pseudoscience, Superstition, and Other Confusions of Our Time* (New York: W. H. Freeman, 1997); Daniel C. Dennett, *Breaking the Spell: Religion as a Natural Phenomenon* (New York: Viking, 2006).

than the doctrines of religion are. The belief that science is the only reliable road to truth certainly lies outside the possible scope of scientific verification, so it is hardly appropriate to demand that religious beliefs be scientifically testable (or falsifiable) either.

Furthermore, theologically speaking, if revelation's content were knowable by the objectifying methods of science, it could not function as revelatory in the first place. The God of revelation is not an object to be mastered but a Subject who invites us to be mastered by an infinite love. Even as far as our ordinary interpersonal relationships are concerned, the epistemological assumptions of scientific naturalism are inapplicable, since we come into contact with the reality of other persons not by objectifying them but by being vulnerable to the claims they make on us. All the more so should we anticipate that the reality of a self-revealing personal God would show up in our awareness only to the extent that we allow it to grasp hold of us. We cannot understand and know it if we insist epistemically on making it into an object alongside those of scientific study. We can have an awareness of being addressed by the self-revealing mystery of God, and we can talk about this mystery in a symbolic or metaphorical way as we share the experience with one another. But we cannot subject it to experimental control. To try to do so would be to remove ourselves from any appropriate relationship to it.

Nevertheless, this does not mean that theologians have an excuse to shy away from the question of whether revelation is true. We may approach this concern by first clarifying what it means to be truthful and then asking whether trust in the promising and self-humbling God of Christian faith opens or closes our minds to the truth. Before embarking on such an exercise, however, we would do well to call to mind once again that for Christians revelation comes in the form of a promise. So we are simply not in a position to verify or falsify it in a publicly comprehensible way at present. Only if and when a promise comes to fulfillment could our trust in it be fully vindicated. Theologian Ronald Thiemann points out that any justification of the truth of revelation "has an inevitable eschatological or prospective dimension. The justifiability of one's trust in the truthfulness of a promise is never fully confirmed (or disconfirmed) until the promiser actually fulfills (or fails to fulfill) his/her promise." Until such future confirmation occurs, Thiemann goes on to say, "the promisee must justify trust on the basis of a judgment concerning the character of the promiser."[4]

---

4. Ronald Thiemann, *Revelation and Theology* (Notre Dame: University of Notre Dame Press, 1985), 94.

With these qualifications in mind, however, I shall argue here that in allowing ourselves to be taken captive by the image of God's descent, and by permitting our aspirations to be suffused with hope in the divine promises, Christian believers can have their minds opened up in such a way as to lend enthusiastic support to the pursuit of truth in general and the journey of scientific discovery in particular. In principle, at least, there should be no hesitation whatsoever on the part of the Christian to undertake a life in science. As I shall argue, revelation's proper effect on consciousness is to make it more, not less, open to scientific pursuits.

WHAT IS TRUTH?

In all discussions of theology and science the question of truth is central and unavoidable. But what is truth? Truth, as both traditional philosophy and modern science have understood it, is the correspondence of the mind with *what is*. In a broader sense, however, truth is that which is sought out by one's desire to know.[5] In order for the reader to gain a more palpable sense of what I mean by this statement I shall be employing second-person discourse in the following exercise.

You may be asking "What do you mean by the desire to know?" Or, "Where is the author leading me in this discussion?" But the fact that you are asking these questions is itself direct evidence that *you* have a desire to know. So you can easily understand what I am talking about simply by attending to your own spontaneous acts of raising questions at this very moment. By becoming explicitly aware of your own questioning mind and its cognitional activity you may arrive at an appropriate point to begin understanding formally the meaning of "truth." Notice now, for example, that your desire to know is leading you to perform three distinct cognitional acts: experiencing, understanding, and judging. As you are reading this book you are *experiencing*, or attending to, what I am writing; second, you are trying to *understand* what I am saying; and, third, you are probably reflecting on and criticizing at least some of what I have been telling you. That is, you are being invited to make a *judgment* as to whether what I am saying is true or false.

You cannot help engaging in these three cognitional acts. This is because at the core of your consciousness lie three corresponding imperatives whose

---

5. Here I shall be following Bernard Lonergan's well-known theory of knowledge; see especially his essay "Cognitional Structure," in *Collection*, ed. F. E. Crowe, S.J. (New York: Herder & Herder, 1967), 221-39.

instructions you can never fully escape. These imperatives, along with the cognitional acts that respond to them are as follows:

1. Be attentive!   →   Experience
2. Be intelligent! →   Understanding
3. Be critical!   →   Judgment

A fourth set (which we need not examine here) is:

4. Be responsible! →   Decision

The imperatives themselves are issued by your desire to know. And so, you can understand "truth" as the name of the lofty goal being sought by the desire to know. If you are asking me at this moment "Are you sure?" you are raising this question only because *you* have a desire to know the truth. You have just caught yourself once again in the act of seeking correct understanding, so the evidence that you have a desire to know and that its goal is truth is directly accessible to you.[6] You cannot deny this without proving the point.

Notice, therefore, that there is more to your mental functioning than seeing or understanding. After all, you can see without understanding, and you can be bright without being right. You can have sight without insight, and insight without truth. Scientists already know this. They implicitly realize that they can approximate truth not just by seeing and thinking but also by being critical of what they understand. Truth is arrived at incrementally not only by looking or understanding but also by obeying the imperative to be critically reflective. Theories and hypotheses have to be tested constantly. Likewise, you can arrive at truth formally only in the distinct cognitional act of *judgment*.

Truth, therefore, can be understood as the objective or goal of your desire to know.[7] Being a lover of truth means, quite simply, *being faithful to your desire to know*. This, in turn, requires that you submit habitually to *all* the imperatives of your mind. These imperatives—to be attentive, intelligent, and critical—emanate from a mysterious font flowing from the very heart of your being. This source is fittingly called the desire to know, and there is no point in trying to dam up its flow. If you are asking "Why not?" then that question itself is already a sign that the dam has burst. The best you can do, if you hope to satisfy your appetite for true knowledge, is to find ways to facilitate its restless search for what is really real. I shall be asking the reader, therefore, to

---

6. Ibid.

7. What I have just summarized is Bernard Lonergan's understanding as expressed in his major work *Insight: A Study of Human Understanding*, 3rd ed. (New York: Philosophical Library, 1970), and in "Cognitional Structure."

reflect on whether faith in the God of Christian revelation functions to liberate the desire to know, or instead suppresses its native longing for *what is*.

If you are a lover of truth, after all, you will do whatever it takes to promote the interests of your desire to know. However, this will not be easy, since you have other desires that also demand satisfaction and whose special pleading is so strong at times that you may barely feel the presence of your desire to know. The longing for physical pleasure, for acceptance by a particular social group, for power and control, and even for meaning—all of which are part of normal human functioning—can at times muffle the imperatives of your mind and silence the desire to know. These other longings will be quite content to rest in illusions as long as they remain out of touch with your desire to know. It is only your desire to know and the imperatives it issues that can be trusted to bring you into touch with truth. So ideally you will look for ways to set this desire free. You cannot remove your other desires, of course, and it is harmful to suppress them, but perhaps their dynamism can be relativized and placed in the service of your desire to know. In any case, it is only when all your longings are in league with, and subordinated to, your desire to know that the truth can come into view and set you free.

REVELATION, SCIENCE, AND THE DESIRE TO KNOW

What I propose, then, is that the trust that awakens in you when you allow yourself to be taken hold of by the revelatory image of God's descent and promise can function to liberate and promote the interests of your desire to know. If this claim turns out to be a reasonable one, then allowing yourself to be moved deeply by revelation will have satisfied the fundamental criterion of truth. Once again, the fundamental criterion of truth, as I am understanding it in accordance with Bernard Lonergan's carefully composed theory of knowledge, is fidelity to the desire to know.[8] If you are careful to listen to the imperatives of your mind, and if you decide to let your desire to know run free to pursue its goal, regardless of where it leads you, you may trust that you are on the way to truth—and freedom.

You will eventually realize, of course, that truth can never be fully possessed because in a sense it possesses you. Indeed, it is only because you have already allowed yourself (partially and tentatively) to be embraced by truth that your desire to know has been awakened in the first place. Even though most sci-

---

8. As I have mentioned already, I am indebted throughout this chapter to Bernard Lonergan's ideas even where my terminology and applications of his theory of knowledge are not precisely his own.

entists fail to notice it, it is only because they too have made a tacit, deeply personal act of surrender to truth as a supreme value that they are properly motivated to do science at all. This is the kind of surrender that Einstein himself considered essential to science. Apart from a deeply personal, indeed "religious," devotion to truth, science is lame. Even the scientific naturalist has in effect bowed down before truth, although the naturalist's attempt to bottle and cap it by scientific method alone is a retreat from the surrender required by the fullness of truth. Anyway, it is your intuitive sense that truth is good (and also beautiful) and that pursuing it can give depth, meaning, and joy to your life that leads you to respect and follow your desire to know. Ideally, therefore, you will look for ways to liberate and strengthen the imperatives of your mind so that they do not get silenced in the cacophony of other desires in which your conscious life can become entangled.

LIBERATING THE DESIRE TO KNOW

The following reflections, therefore, are not intended as a public exhibition of the truth of revelation in conformity with the ordinary scientific requirements of gaining cognitional control of a specific set of objects. Such an impersonal and detached mode of justification would cause you to lose sight of the subject matter of revelation, which can seize you and empower you to seek truth. The kind of epistemic justification that fits the experience of revelation can only be an indirect one. Only after surrendering to the claims of revelation, and at the same time having become fully aware of your desire to know, are you in a position to assess the truth status of your faith. What I propose, then, is that if you have been deeply moved by the symbolic disclosure of ultimate reality in Christ, you might ask now whether the trust inspired by this revelatory experience frustrates or supports your own desire to know. If the experience of faith leads you to suppress the deepest and most trustworthy of all your longings, then honesty requires that you let go of that faith, as the naturalist will rightly advise you. But if your being formed and informed by the revelatory image serves to liberate your desire to know, then it has passed the test of fulfilling the fundamental criterion of truth.

It goes without saying that you cannot undertake this deeply personal exercise without having first been caught up in the circle of faith in the Christian revelation. You cannot decide the question of revelation's truth status from a neutral, dispassionate perspective such as that of scientific method, since faith, by definition, is a matter of ultimate concern.[9] But this does not mean that the

---

9. Paul Tillich, *Dynamics of Faith* (New York: Harper Torchbooks, 1958), 1.

only option left is a fideist one in which the demands of your reason are dismissed as irrelevant, and faith is accepted as an irrational leap of the mind. Only after you have been grasped by the power of the revelatory image and responded to it with faith, hope, and love are you in a position to ask about its truth status, that is, whether your belief is consistent with reason. You have to try it on before you can determine whether it is a good fit epistemologically, but you may still fruitfully ask whether your trust in revelation's God supports your desire to know.

At this point, of course, the naturalist will complain that it is irrational to make a leap of faith in the first place.[10] But you may fittingly reply that naturalism itself is no less of a leap, since it wagers that "nature is all there is" without ever having proven this to be the case, scientifically speaking. So naturalists themselves would do well to undertake the same exercise I am laying out here. They should ask themselves whether their own leap of faith or their trust in the naturalistic worldview is completely consistent with the full liberation of their own desire to know.[11]

At any rate, it is only after having already been drawn into the circle of faith that you may ask whether your mind's being influenced by this faith supports or sabotages the native interests of your desire to know. Each believer must undertake this examination internally, so the following can only be a sketch of the steps one may take in responding to the question of whether faith in the Christian revelation, along with the hope in the future that it inspires, can justifiably be called truthful in the age of science. I do not want to give the impression that the exercise is easy, or that anyone has carried it out with complete success, or that darkness can ever be dispelled from the journey of faith. The truth of revelation can never be arrived at in such a way that trust in it does not have to be renewed each day. Whether trust in revelation is consistent with full fidelity to the desire to know, and is therefore reasonable, cannot be settled irreversibly in the dispassionate manner of scientific argumentation. The following, therefore, is nothing more than a way of showing that in principle a life of faith can be one that supports the desire to know and the demands of reason, and hence it cannot be logically refuted by the gratuitous claims of scientific naturalism.

---

10. See the famous declaration of W. K Clifford: "It is wrong always, everywhere, and for anyone, to believe anything upon insufficient evidence." This statement was criticized by William James in his important essay "The Will to Believe," in *The Will to Believe, and Other Essays in Popular Philosophy* (New York: Longmans, Green, 1931).

11. In my book *Is Nature Enough? Meaning and Truth in the Age of Science* (Cambridge: Cambridge University Press, 2006) I have argued at much greater length that naturalism is not only inconsistent with the desire to know but also that it implicitly subverts our native longing for truth.

TRUTH AND THE DESCENT OF GOD

Trust in a God of infinite, selfless, and all-encompassing love can be said to be truthful inasmuch as it motivates one to remove obstacles to the free flow of the desire to know. But probably nothing cripples and sidelines the desire to know more effectively than the unrestrained exercise of the will to power, which in its most destructive extreme amounts to the will to absolute control. Therefore, in faith's appropriation of the revelatory image of God's humility, the mind of the recipient of revelation is invited to release itself from bondage to the instinctive need for absolute control and to surrender itself more fully to the desire to know. It is in this way that trust in revelation can serve the cause of truth.

Such is the argument in a nutshell, but it needs unpacking. Since doing so would require at least another book, I can provide only an outline here.[12] Let me begin by expressing agreement with those philosophers and psychologists who observe that the will to power is in itself a natural instinct and that repression of it is harmful. Psychologically the development of a strong ego, and with it a sense of self-esteem, is an essential part of normal development. A sense of empowerment is necessary in both individual and social existence, and repression of our need for it can be debilitating. What I am talking about here is the *unrestrained* instinct to mastery or control that arises when the will to power becomes detached from other longings such as the need to belong, the need to be loved, the drive to understand, and the desire to know. Almost any dimension of human culture, especially the political, but also the intellectual, scientific, and religious domains, can be taken hostage by the impulse to control. When the untamed will to power inhabits the sociopolitical arena, great evils ensue: pogroms, holocausts, environmental disasters, unprovoked wars, and so on. In the religious realm the will to mastery sometimes takes the form of an obsession with certitude, embodying itself in scriptural or doctrinal literalism that splits the world into true believers and nonbelievers, a divisiveness that can also lead to inquisitions, destruction, and death.

An unrestrained will to power can sometimes take over the domain of scientific understanding also. For example, in some academic and research institutes where science is given an important role—and where one would expect the desire to know to be dominant—the impulse to control can sometimes silently and seductively creep in. Often completely unnoticed, it begins to subordinate to itself the more sensitive and vulnerable desire to know. Specifi-

---

12. I have attempted a book length treatment of this issue in *Religion and Self-Acceptance* (New York: Paulist, 1976).

cally the will to mastery typically takes the form of an absolutist *reductionism*, which I earlier defined as the suppression of layered explanation. The declaration that life can be understood only in terms of chemistry, for example, is not knowledge at all but a manipulative suppression of the wide empiricism needed to gain a nuanced understanding of the world.

I am not speaking here, however, of science as such. Science is an authentic and fruitful unfolding of the pure desire to know. Moreover, science appropriately employs a reductive method in order to focus its inquiry on a manageable set of data. But at times the methodological reduction characteristic of scientific method can be taken hostage by the will to mastery. Whenever this takeover occurs, the world shrinks in size and complexity. Thought loses touch with the native wildness and elusiveness of the knowable world, and Occam's razor becomes a weapon of destruction. Once it has been taken prisoner by scientism, science has the goal no longer of expanding knowledge but instead of putting arbitrary limits on what can be taken as true or real. And so nature gets fictionalized as manipulable machinery, or as a set of mindless particles-in-motion, subject in principle to complete scientific and technological control. Here the humble and inherently vulnerable desire to know has been shoved aside by an aggressive impulse to subject all of reality to complete scientific hegemony.

The outcome of this assault is a new, quasi-religious dogmatism known as materialism (or physicalism), the belief that the physical world, as made intelligible by scientific method, is really all there is. Such a metaphysical outlook may seem innocuous and even cognitionally indispensable at first sight, but Carolyn Merchant has argued persuasively that it has led directly to the "death of nature," now made agonizingly evident in the world's ecological predicament.[13] Similarly Michael Polanyi has shown that modern materialism's implicit objectification of nature leads logically to a nihilistic worldview in which persons, values, and even life become virtually absent from the universe. Concrete evidence of the incursion of the materialist and reductionist habit of mind into social and political agendas can be found in the last century's unprecedented disregard of persons, most glaringly in the manipulation—and even annihilation—of humans by ruthless dictators. Materialism cannot help but provide intellectual support for the suspicion that life has no intrinsic value since its ontological status is that of being degradable into lifeless stuff.

According to Polanyi, a "moral inversion" began to take place in the late

---

13. Carolyn Merchant, *The Death of Nature: Women, Ecology, and the Scientific Revolution* (San Francisco: Harper & Row, 1980).

modern world whereby the natural ethical instincts of countless people in the world were enlisted to execute intellectual culture's new imperative to objectify, and hence depersonalize, the world.[14] It even came to be thought of as ethically wrong, as Jacques Monod later declared, to violate the "postulate of objectivity," according to which everything real must be subjected to scientific method.[15] Consequently, the venerable idea that subjectivity is real became increasingly taboo in the age of science.[16] Once it has been made subservient to an exclusively materialistic understanding, subjectivity can only be thought of as less real than the lifeless "matter" of which it is thought to be composed. Having been transferred epistemologically to the realm of fully objectifiable being, persons become easier prey than ever to the manipulative will of social engineers.

Fortunately, even die-hard materialists are usually inconsistent enough not to draw such dire conclusions in their own ethical lives. But the subtle subordination of the desire to know to a reductionist will to control remains a significant element in the atmosphere of contemporary scientific and intellectual culture. Even though postmodern criticism has exposed the manipulative will to control that energizes naturalistic reductionism, the latter continues to exercise enormous sociocultural influence. So it is important, in the interest of truthfulness, to be aware of how destructive a purely objectifying method of investigation can become whenever it becomes isolated from the pure desire to know and is placed in the service of less noble instincts.

IMPLICATIONS OF REVELATION

In light of these considerations, therefore, I suggest that the truth of revelation can be affirmed to the extent that it allows us to trust that the very Ground of Being is itself humble, self-giving love rather than a will-to-control. To be seized by an infinite love means, by definition, to have renounced any unrestrained will to mastery. If in faith we experience the universe's own creative ground as calling things into being by a loving "letting be," rather than by a crudely causal exercise of force, then the insight may occur to us, especially in our dealings with other free beings—and to an extent with all

14. Michael Polanyi, *Personal Knowledge: Towards a Post-Critical Philosophy* (New York: Harper Torchbooks, 1964), 231-37.

15. Jacques Monod, *Chance and Necessity: An Essay on the Natural Philosophy of Modern Biology*, trans. Austryn Wainhouse (New York: Knopf, 1971), 175-80.

16. B. Alan Wallace, *The Taboo of Subjectivity: Toward a New Science of Consciousness* (New York: Oxford University Press, 2000).

other living beings as well—that there is something inherently unmanipulable about life and persons. If we think of life and love as real, and if we refuse to reduce them without remainder to what is chemically elemental, the universe need no longer be thought of as a deterministic machine blindly subject to a series of mechanical causes arising out of the dead cosmic past. Instead, the cosmos will be apprehended as a story of emergent freedom invited into being by the indeterminate promise of a God encountered on the brink of an always receding future horizon.

However, it is not the case that the selfless God who has become shockingly manifest in the crucified Christ is now in every sense powerless. Rather, power has been radically, decisively, and everlastingly redefined, not eliminated. If power means the capacity to influence, to produce effects and make a significant difference, then Christians will still profess belief in an omnipotent and almighty God. In the new dispensation, however, power means something different from what it did before. The image of the divine descent breaks through the loveless world that is still in bondage to a demonic exercise of force, including the bondage that accompanies all loveless images of the deity. Revelation proclaims that henceforth any true notion of power and authority has been transfigured by an endlessly self-giving love. In the very event of exposing the emptiness of an unrestrained will to mastery, the kenotic image of God's descent releases the desire to know, at least in principle, from its chief rival, the runaway will to power.

RELEASE FROM SELF-DECEPTION

Another obstacle to the liberation of the desire to know is our tendency to deceive ourselves. Ultimately self-deception is caused by fear, especially fear of disapproval. Hence, whatever takes away unnecessary fear of disapproval can serve to render self-deception pointless. Perfect love casts out fear, so to the extent that we allow ourselves to be encircled by selfless love we can be delivered of the anxiety that causes us to deceive ourselves. Self-deception arises when we try hard to conform to an ideal and then find ourselves unacceptably falling short. The capacity for self-deception, as Herbert Fingarette has noticed, is possible only if an individual has first aspired to what is considered good.[17] An individual who has not yet been attracted to an ideal or conformed to the requirements of social existence will not be subject to self-deception. For example, a sociopath or a person who is consciously and willfully com-

---

17. Herbert Fingarette, *Self-Deception* (Berkeley: University of California Press, 2000), 1, 139-44.

mitted to doing evil has no need for self-deception. Instead it is the broad swath of humans who have found ideals to live by, but who at the same time are in some way unable or unwilling to live up to them, who are prey to self-deception. If we think of our self-worth as contingent upon living up to the standards associated with what we idealize, then any shortcomings we have may plunge us into either despair or self-deception.

Trust in the self-humbling love of God disclosed in the cross of Christ, it would appear, can in principle have the effect of deposing the tyrannical gods of everyday consciousness, whose images have served to make us hide from ourselves. Sadly, even our customary psychic representations of the biblical God can tend at times to cause rather than alleviate pathological anxiety, thus shoring up self-deception and, in doing so, frustrating the desire to know. As the religious experiences of St. Paul and Martin Luther illustrate in an especially vivid way, there are ways of thinking about God that function to keep one in fear and self-deception. Indeed, some of the most interesting forms of modern atheism, including those of scientific naturalism, are in effect expressions of protest against the enshrinement in psyche and society of images of God whose function is to sanction forms of power that promote fear and self-deception. Insofar as these images keep the desire to know from reaching the truth about oneself they are illusory and must be rejected as inconsistent with the fundamental criterion of truth.[18]

Among my friends and acquaintances there are many former Christians who would now call themselves scientific naturalists. I am convinced that most of them have abandoned theism because they have found the idea of a "personal" God intolerable. And they have found it intolerable because it seems to them to promote typically a kind of religiosity that keeps them from a full acceptance of themselves. Often, therefore, their movement toward atheism, whether gradual or headlong, has been accompanied by a sense of relief and inner liberation. Having to settle for the impersonal universe of naturalism seems to them to be worth the price of being delivered of the intolerable burden of self-deception and self-hatred that had accompanied their former religious experience.[19]

In thinking of God as personal, biblical religions have taken a great risk, since almost any set of attributes, healthy or not, can be projected onto persons. It is partly for this reason that I have taken pains to present as founda-

---

18. I cannot develop the point here, but it is relevant to ask to what extent exclusively patriarchal images of God promote self-deception and in doing so frustrate the desire to know.

19. There are, of course, other and much more complicated reasons why atheism is attractive to many people, but there are many written accounts by atheists and agnostics during the last two centuries that support the point I am making here.

tional to Christian faith the image of a self-humbling God who identifies with the broken, the anxious, the oppressed, and the unaccepted, a God who has nothing to do with potentates who survive only by virtue of the fear their subjects have for them.

The image of God revealed in the Christ of the Gospels is that of vulnerable, self-giving love. Yet it is not at all an image of weakness and impotence, since it has the power to remove the stigma of shame that leads to self-deception. It has the strength to restore to estranged and forgotten people a sense of their intrinsic value in a more profound and enduring way than any dictatorial, political, or mechanical manipulations could ever accomplish. What stands out in the Gospels is that the divine love incarnate in Christ is experienced as a power of renewal, giving people an unprecedented sense of their self-worth. A God who descends into the domain of what has been lost, who identifies with the estranged and forgotten, can be experienced as embracing also those portions of our own selfhood which, for the sake of gaining approval, we may have pushed out of explicit consciousness. God's descent, symbolized powerfully in the Christian imagery of Jesus searching for what is lost, even to the point of his "descent into hell"—into the realm of the relationless, the hopeless, and the dead—can penetrate to the depths of the self also, as the experience of St. Paul testifies. If God can be experienced as embracing aspects of myself and my life story that I find unacceptable, trust in such a God can empower me to accept them also.

Thus, faith in God's unbounded love and trust in the promise of everlasting fidelity can, at least in principle, function to liberate the desire to know from the self-deception that stands between itself and truth. The testimony of the great religious and philosophical traditions, as well as wise people throughout the ages, is that self-deception is the most difficult obstacle to being consistently truthful. But Herbert Fingarette is especially puzzled by the fact that philosophical insights alone have done so little to understand or remove self-deception:

> Were a portrait of man to be drawn, one in which there would be highlighted whatever is most human, be it noble or ignoble, we should surely place well in the foreground man's enormous capacity for self-deception. The task of representing this most intimate, secret gesture would not be much easier were we to turn to what philosophers have said. Philosophical attempts to elucidate the concept of self-deception have ended in paradox or in loss from sight of the elusive phenomenon itself.[20]

---

20. Fingarette, *Self-Deception*, 1.

Can a posture of religious trust perhaps be more successful than philo-sophical reasoning in the treatment of self-deception? Whatever the motivat-ing factors for self-deception may be, trusting that the ultimate environment of our lives is a humble, self-limiting love may allow the desire to know to break through the walls behind which fear and self-deception have concealed it. If we understand God fundamentally as limitless love, there should be no reason to hide from ourselves, since an infinite, and hence unconditional, love embraces us in spite of what seems unacceptable to us. God has loved us "while we were still in our sin" (Rom 5:8), and God is "kind to the ungrateful and the selfish" (Luke 6:35). Were we to become fully imbued with trust in such a God, there would be no need to suppress awareness of what we had previously taken to be unapprovable about ourselves. We would be released from the energy-sapping tendency to censor portions of ourselves. At least in an imperfect way the armor of self-deception could be laid aside, and the desire to know could then be set free from the distortive cage in which it had been locked away. Thus, the image of God's descent as revealed in the picture of Jesus' embrace of the unaccepted, when applied to our own inner life as well as our social life, is in the service of the desire to know. By trusting in such a God we are able to fulfill the fundamental criterion of truth.

TRUTH AND THE GOD OF PROMISE

Following Lonergan's thought, I am taking the fundamental criterion of truth to be fidelity to the desire to know. Accordingly, faith in revelation could be called truthful if it serves to promote the interests of this desire. By definition, as the exercise prescribed above has attempted to demonstrate, the desire to know can never be satisfied with deception and illusion. So any posture of consciousness that motivates one to be objective about oneself will quite likely also promote objectivity about everything else—even the natural world. Fail-ing to expose the darker side of our own selves to the illuminating and puri-fying light of an infinite love, we can easily go through life misreading other people and even the universe itself as embodiments of the hostility and indif-ference that we have failed to acknowledge in ourselves. Any of us may be tempted at times to interpret nature as unequivocally evil—a "wicked old witch" in the words of biologist George Williams, or "pitiless indifference" in those of Richard Dawkins—even though at worst the world is ambiguous and has been most generous in giving rise to life and human existence. Still, if we happen to be scientifically learned and at the same time believe that "nature is all there is," we might declare that the whole universe is *essentially* lifeless, mindless, and pointless.

This cosmic pessimism still passes as high wisdom in intellectual circles today. But in terms of the relationship of science to faith there is still room to ask whether resignation to tragic "realism" is more conducive to liberating the desire to know than is trust in the revelatory promises of Christian faith. Suppose, for the moment, that we follow the logic of revelation and learn to "see" the world as grounded, first, in a love that is humble enough to let the universe be truly other than its Creator and, second, in an everlasting fidelity that opens the world to an always new future. There is nothing in the discoveries of the natural sciences, I would suggest, that discourages or forbids such a contextualizing of the cosmos. Indeed, the phenomenon of emergence, as I have already pointed out, fits more comfortably into Christianity's anticipatory vision of reality than into the "metaphysics of the past" absolutized by scientific naturalism and cosmic pessimism. When combined with science's new sense of an unfinished universe, biblical hope in a future of new creation may well turn out to be the most realistic stance one can take in an ambiguous world.

By allowing our lives to be informed by a sense of God's covenantal fidelity—a disposition that nearly every book in the Bible begs us to consider—the desire to know, including its scientific leanings, will be enabled to flow much more freely toward its goal than if such trust were absent. A deeply felt conviction that a new future lies open ahead of us and our unfinished universe releases not only the heart to a life of hope but also the mind to a life of limitless exploration. Surely science can thrive within such a hopeful vision of things. Therefore, I conclude that science, along with other kinds of experiencing, understanding, and critical reflection, is not in conflict with, but can actually draw support from, the trust awakened by the promissory aspect of revelation.

Yet how can this be so? Let me begin a very brief response by emphasizing once again that in biblical experience the mystery of God reveals itself as having the character of a promise that opens up the prospect of the emergence of more intense modes of being in the future. Since being is proportionate to consciousness, "more being" means more consciousness. Thus, as Teilhard observes, what we can learn from the still evolving universe, to guide us "through the fogbanks of life," is that we should direct our actions and lives "toward the greater degree of consciousness." If we do so, "we can be certain of sailing in convoy with and making port with the universe." As we look into cosmic history we can see that the world has always allowed for a *further* increase in consciousness—and for Teilhard this means *more being*—up ahead. It still does today, perhaps more obviously than ever. Consequently, ". . . we should use the following as an absolute principle of appraisal in our judgments: 'It is better, no matter what the cost, to be more conscious than less conscious.'" "This principle," Teilhard insists, "is the absolute condition of the world's existence."[21]

In the temporal context of our existence, however, the infinite self-giving mystery we call God opens the universe to "more being" by also being the Absolute Future that carries the cosmos toward itself. Since the universe we inhabit is finite, it could never receive the fullness of the divine infinity in any conclusive instant. God's revelatory self-gift, in other words, cannot fully disclose itself completely in any single moment of the world's unfolding. In its boundless generosity God's self-limiting love always remains partially hidden. This itself is a great gift, since by not becoming fully present now, God opens a future for the cosmos, offering it the opportunity to become new each day as well as to participate in its own creation.

Theologian Wolfhart Pannenberg, therefore, justifiably defines revelation as the "arrival of the future."[22] It is especially in Jesus' resurrection that the world's ultimate future reveals itself under the limiting conditions of the present. "By contemplating Jesus' resurrection, we perceive our own ultimate future," Pannenberg writes. And, most significantly, the aura of incomprehensibility that now surrounds the mystery of the resurrection "means that for the Christian the future is still open and full of possibilities." The divine futurity reveals itself in human history primarily in the mode of promise. And a theology of nature may assume that the same promise that came to Abraham and his descendants also comes even now to meet the cosmos in each moment, opening it up to a future emergence on a scale too grand for any individual life or single epoch to calculate. The God of the Bible who "goes before" the people of Israel also goes before the evolving universe.

Of course, in order to experience the natural world as radically open to the future we must first put on the virtue of *hope*. To the naturalist, hope might seem unrealistic, but, unlike the will to certitude, hope can live comfortably in an unfinished and ambiguous universe, and in this respect it matches up more fittingly with what science has uncovered than do the dogmatic constraints of scientific naturalism. An unfinished universe, after all, cannot be fully intelligible here and now, so the desire to know, whose very nature is to seek a *fullness* of intelligibility and truth, must attach itself to a trustful hope that can wait in patience. Unlike the much greedier will to power, which insists on absolute certainty here and now, the desire to know is willing to wait for *all*

---

21. Pierre Teilhard de Chardin, *How I Believe*, trans. René Hague (New York: Harper & Row, 1969), 35.

22. Wolfhart Pannenberg, *Faith and Reality*, trans. John Maxwell (Philadelphia: Westminster, 1977), 58-59.

the data to come in. It does not insist on complete possession of clarity here and now. We may conclude, then, that the revelatory image of God's futurity and the hope it inspires are fully supportive of the desire to know that underlies rational and scientific pursuits.

"Our world contains within itself a mysterious promise of the future, implicit in its natural evolution," says Teilhard, "that is the final assertion of the scientist as he closes his eyes, heavy and weary from having seen so much that he could not express."[23] A leaning toward the future is a fundamental feature of nature as such. This tendency finds its fullest flowering in religious hope for a final fulfillment of all things. Important as it is to scientific understanding, tracing the causes of present phenomena back into the remote past or into the elemental levels of cosmic chemistry, leads only toward the incoherence and unintelligibility of mere multiplicity if we go back far enough.[24] Only by looking *from* the past *toward* the future fulfillment of the cosmic process can we expect intelligibility to begin showing up. Complete intelligibility, therefore, can only coincide with the world's ultimate encounter with its Absolute Future, a goal that we can approach here and now only by cultivating the virtue of hope.[25] It is only on the wings of hope that the desire to know can be set completely free to understand and know the universe.

SUGGESTIONS FOR FURTHER READING AND STUDY

Haught, John F. *Is Nature Enough? Meaning and Truth in the Age of Science.* Cambridge: Cambridge University Press, 2006.

Lonergan, Bernard. "Cognitional Structure." In *Collection,* edited by F. E. Crowe, S.J., 221-39. New York: Herder & Herder, 1967.

Mooney, Christopher. *Theology and Scientific Knowledge.* Notre Dame: University of Notre Dame Press, 1996.

Ogden, Schubert. *The Reality of God and Other Essays.* New York: Harper & Row, 1977.

---

23. Pierre Teilhard de Chardin, *Writings in Time of War,* trans. René Hague (New York: Harper & Row, 1968), 55-56.

24. A point that Teilhard de Chardin makes throughout his works.

25. Karl Rahner, S.J., *Theological Investigations,* vol. 6, trans. Karl and Boniface Kruger (Baltimore: Helicon, 1969), 59-68.

Selected Bibliography on Science and Christianity

Atkins, Peter W. 1992. *Creation Revisited*. New York: W. H. Freeman.

———. 1994. *The 2nd Law: Energy, Chaos, and Form*. New York: Scientific American Books.

Atran, Scott. 2002. *In Gods We Trust: The Evolutionary Landscape of Religion*. New York: Oxford University Press.

Barbour, Ian G. 1968. "Five Ways to Read Teilhard." *The Teilhard Review* 3:3-20.

———. 1997. *Religion and Science: Historical and Contemporary Issues*. San Francisco: HarperSanFrancisco.

Barrow, John, and Frank Tipler. 1986. *The Anthropic Cosmological Principle*. Oxford: Clarendon Press.

Behe, Michael J. 1996. *Darwin's Black Box: The Biochemical Challenge to Evolution*. New York: Free Press.

Benz, Ernst. 1966. *Evolution and Christian Hope: Man's Concept of the Future from the Early Fathers to Teilhard de Chardin*. Translated by Heinz G. Frank. Garden City, N.Y.: Doubleday Anchor Books.

Berger, Peter L. 1969. *A Rumor of Angels: Modern Society and the Rediscovery of the Supernatural*. Garden City, N.Y.: Doubleday.

———. 1990. *The Sacred Canopy: Elements of a Sociological Theory of Religion*. Garden City, N.Y.: Anchor Books.

Bergson, Henri. 1983. *Creative Evolution*. Translated by Arthur Mitchell. Lanham, Md.: University Press of America.

Bowler, Peter. 2001. *Reconciling Science and Religion: The Debate in Early Twentieth-Century Britain*. Chicago: University of Chicago Press.

Boyer, Pascal. 2001. *Religion Explained: The Evolutionary Origins of Religious Thought*. New York: Basic Books.

Brockman, John. 1995. *The Third Culture*. New York: Touchstone.

Buckley, Michael J. 1987. *At the Origins of Modern Atheism*. New Haven: Yale University Press.

Bulst, Werner. 1965. *Revelation*. Translated by Bruce Vawter. New York: Sheed & Ward.

Bultmann, Rudolf. 1961. "New Testament and Mythology." In *Kerygma and Myth*, edited by Hans Werner Bartsch. Translated by Reginald Fuller. New York: Harper Torchbooks.

Burkert, Walter. 1996. *Creation of the Sacred: Tracks of Biology in Early Religions*. Cambridge, Mass.: Harvard University Press.

Burtt, E. A. 1954. *The Metaphysical Foundations of Modern Science*. Garden City, N.Y.: Doubleday Anchor Books.

Camus, Albert. 1955. *The Myth of Sisyphus, and Other Essays*. Translated by Justin O'Brien. New York: Knopf.

Chaitin, Gregory J. 2003. "Randomness and Mathematical Proof." In *From Complexity to Life: On the Emergence of Life and Meaning*, edited by Niels Gregersen. New York: Oxford University Press.

Chalmers, David. 1995. "Facing Up to the Problem of Consciousness." *Journal of Consciousness Studies* 2:200-219.

Churchland, Paul M. 1995. *The Engine of Reason, the Seat of the Soul: A Philosophical Journey into the Brain*. Cambridge, Mass.: MIT Press.

Cobb, John B., Jr., and Charles Birch. 1988. *The Liberation of Life: From the Cell to the Community*. Cambridge: Cambridge University Press.

Cobb, John B., Jr., and David Griffin. 1976. *Process Theology: An Introductory Exposition* Philadelphia: Westminster.

Conway Morris, Simon. 2003. *Life's Solution: Inevitable Humans in a Lonely Universe*. New York: Cambridge University Press.

Crews, Frederick. 2001. "Saving Us from Darwin." *New York Review of Books*, Part I: October 4; Part II: October 18.

Crick, Francis H. C. 1966. *Of Molecules and Men*. Seattle: University of Washington Press.

Cziko, Gary. 1995. *Without Miracles: Universal Selection Theory and the Second Darwinian Revolution*. Cambridge. Mass.: MIT Press.

Darwin, Charles. 1958 [1892]. *The Autobiography of Charles Darwin, 1809-1882: With Original Omissions Restored*. Edited by Nora Barlow. New York: Harcourt.

Davies, Paul. 1983. *God and the New Physics*. New York: Simon & Schuster.

———. 1993. *The Mind of God: The Scientific Basis for a Rational World*. New York: Simon & Schuster.

Dawe, Donald G. 1963. *The Form of a Servant: A Historical Analysis of the Kenotic Motif*. Philadelphia: Westminster.

Dawkins, Richard. 1986. *The Blind Watchmaker*. New York: W. W. Norton.

———. 1995. *River Out of Eden: A Darwinian View of Life*. New York: Basic Books.

———. 1996. *Climbing Mount Improbable*. New York: W. W. Norton.

Delio, Ilia. 2001. *Simply Bonaventure: An Introduction to His Life, Thought, and Writings*. New York: New City Press.

———. 2005. *The Humility of God*. Cincinnati: St. Anthony Messenger Press.

Dembski, William A. 1999. *Intelligent Design: The Bridge between Science and Theology*. Downers Grove, Ill.: InterVarsity.

———, ed. 1998. *Mere Creation: Science, Faith and Intelligent Design*. Downers Grove, IL: InterVarsity Press.

Dennett, Daniel C. 1991. *Consciousness Explained*. New York: Little, Brown.

———. 1995. *Darwin's Dangerous Idea: Evolution and the Meaning of Life*. New York: Simon & Schuster.

———. 2006. *Breaking the Spell: Religion as a Natural Phenomenon*. New York: Viking.

Draper, John William. 1898. *History of the Conflict between Religion and Science*. New York: D. Appleton.

Einstein, Albert. 1994. *Ideas and Opinions.* New York: Modern Library.

Ferris, Timothy. 1988. *Coming of Age in the Milky Way.* New York: Doubleday.

Fingarette, Herbert. 2000. *Self-Deception.* Berkeley: University of California Press.

Flanagan, Owen. 2002. *The Problem of the Soul: Two Visions of Mind and How to Reconcile Them.* New York: Basic Books.

Geisler, Norman L. 1982. *Miracles and Modern Thought.* Grand Rapids: Zondervan.

Geisler, Norman L., and J. Kerby Anderson. 1993. "Origin Science." In *Religion and the Natural Sciences,* edited by James E. Huchingson, 199-204. New York: Harcourt, Brace, Jovanovich.

Gillispie, Charles Coulston. 1996. *Genesis and Geology: A Study in the Relations of Scientific Thought, Natural Theology, and Social Opinion in Great Britain, 1790-1850.* Cambridge, Mass.: Harvard University Press.

Goldberg, Stephen. 1999. *Seduced by Science: How American Religion Has Lost Its Way.* New York: New York University Press.

Goodenough, Ursula. 1998. *The Sacred Depths of Nature.* New York: Oxford University Press.

Gould, Stephen Jay. 1977. *Ever Since Darwin: Reflections in Natural History.* New York: W. W. Norton.

———. 1989. *Wonderful Life: The Burgess Shale and the Nature of History.* New York: W. W. Norton.

Gould, S. J., and R. C. Lewontin, 1979. "The Spandrels of San Marco and the Panglossian Paradigm: A Critique of the Adaptationist Programme." *Proceedings of the Royal Society of London,* Series B, 205, No. 1161: 581-98.

Greeley, Andrew M. 1985. *Unsecular Man: The Persistence of Religion.* New York: Schocken Books.

Gribbin, John. 1993. *In the Beginning: After COBE and before the Big Bang.* New York: Little, Brown.

Griffin, David Ray. 2000. *Religion and Scientific Naturalism: Overcoming the Conflicts.* Albany: State University of New York Press.

Grumett, David. 2005. *Teilhard de Chardin: Theology, Humanity and Cosmos.* Studies in Philosophical Theology 29. Leuven and Dudley, Mass.: Peeters.

Hallman, Joseph. 1991. *The Descent of God: Divine Suffering in History and Theology.* Minneapolis: Fortress.

Hardwick, Charley. 1996. *Events of Grace : Naturalism, Existentialism, and Theology.* New York: Cambridge University Press.

Harris, Sam. 2004. *The End of Faith: Religion, Terror, and the Future of Reason.* New York: W. W. Norton.

Hartshorne, Charles. 1941. *Man's Vision of God.* Chicago and New York: Willett, Clark.

———. 1948. *The Divine Relativity.* New Haven: Yale University Press.

Haught, John F. 1976. *Religion and Self-Acceptance.* New York: Paulist.

———. 1986. *What Is God? How to Think about the Divine.* Mahwah, N.J., and New York: Paulist.

———. 1993. *Mystery and Promise: A Theology of Revelation.* Collegeville: Liturgical Press.

———. 1993. *The Promise of Nature*. Mahwah, N.J., and New York: Paulist.

———. 1995. *Science and Religion: From Conflict to Conversation*. Mahwah, N.J., and New York: Paulist.

———. 2000. *God after Darwin: A Theology of Evolution*. Boulder, Colo.: Westview.

———. 2003. *Deeper than Darwin: The Prospect for Religion in the Age of Evolution*. Boulder, Colo.: Westview.

———. 2005. "Darwin, Design and the Promise of Nature." *Science and Christian Faith* 17:5-20.

———. 2005. "What If Theology Took Evolution Seriously." *New Theology Review* 18:10-20.

———. 2006. *Is Nature Enough? Meaning and Truth in the Age of Science*. Cambridge: Cambridge University Press.

Hawking, Stephen W. 1988. *A Brief History of Time*. New York: Bantam Books.

Henderson, Lawrence J. 1913. *Fitness of the Environment: An Inquiry into the Biological Significance of the Properties of Matter*. New York: Macmillan.

Hick, John. 1978. *Evil and the God of Love*. Rev. ed. New York: Harper & Row.

Holland, John H. 1999. *Emergence: From Chaos to Order*. New York: Perseus Books.

Hull, David. 1992. "The God of the Galapagos." *Nature* 352:485-86.

Humphrey, Nicholas. 1996. *Leaps of Faith: Science, Miracles and the Search for Supernatural Consolation*. New York: Basic Books.

Jaki, Stanley L. 1992. *Universe and Creed*. Milwaukee: Marquette University Press.

James, William. 1964 [1907]. *Pragmatism*. Cleveland: Meridian Books.

Jastrow, Robert. 1978. *God and the Astronomers*. New York: W. W. Norton.

Jeans, James. 1948. *The Mysterious Universe*. Rev. ed. New York: Macmillan.

John Paul II, Pope. 1979. "Address to the Pontifical Academy of Sciences." *Origins, CNS Documentary Service* 9, 24.

———. 1988. "Letter to the Reverend George V. Coyne, S.J., Director of the Vatican Observatory." *Origins* 18, 23.

———. 1998. Encyclical letter *Fides et Ratio*, http://www.vatican.va/holy_father/john _paul_ii/encyclicals/documents/hf_jp-ii_enc_15101998_fides-et-raio_en.html.

Johnson, Phillip E. 1999. *The Wedge of Truth: Splitting the Foundations of Naturalism*. Downers Grove, Ill.: InterVarsity.

Johnson, Stephen. 2001. *Emergence: The Connected Lives of Ants, Brains, Cities, and Software*. New York: Touchstone.

Jonas, Hans. 1966. *The Phenomenon of Life*. New York: Harper & Row.

Jüngel, Eberhard. 1976. *The Doctrine of the Trinity: God's Being Is in Becoming*. Translated by Scottish Academic Press Ltd. Grand Rapids: Eerdmans.

Kelly, Anthony. 2006. *Eschatology and Hope*. Theology in Global Perspective. Maryknoll, N.Y.: Orbis Books.

King, Thomas M. 1983. "Teilhard and Piltdown." In *Teilhard and the Unity of Knowledge*, edited by Thomas M. King, S.J., and James Salmon, S.J., 159-69. New York: Paulist.

———. 2005. *Teilhard's Mass: Approaches to "The Mass on the World."* New York: Paulist.

Küng, Hans. 1984. *Eternal Life: Life after Death as a Medical, Philosophical and Theological Problem.* Translated by Edward Quinn. Garden City, N.Y.: Doubleday.

Lackey, Douglas. 1993. "The Big Bang and the Cosmological Argument." In *Religion and the Natural Sciences,* edited by James E. Huchingson, 190-95. New York: Harcourt, Brace, Jovanovich.

Lacouture, Jean. 1996. *Jesuits: A Multibiography.* London: Harvill.

Larson, E. J., and L. Witham. 1999. "Scientists and Religion in America." *Scientific American* 281:88-93.

Lindberg, David C., and Ronald Numbers, eds. 2003. *When Science and Christianity Meet.* Chicago: University of Chicago Press.

Linde, Andrei. 2000. "Inflationary Cosmology and the Question of Teleology." In *Science and Religion in Search of Cosmic Purpose,* edited by John F. Haught, 1-17. Washington, D.C.: Georgetown University Press.

Lonergan, Bernard. 1967. "Cognitional Structure." In *Collection,* edited by F. E. Crowe, S.J., 221-39. New York: Herder & Herder.

———. 1970. *Insight: A Study of Human Understanding.* 3rd ed. New York: Philosophical Library.

Lovejoy, Arthur O. 1965. *The Great Chain of Being: A Study of the History of an Idea.* New York: Harper & Row.

Macquarrie, John. 1978. *The Humility of God.* Philadelphia: Westminster.

Marcel, Gabriel. 1949. *Being and Having.* Westminster: Dacre Press.

Mayr, Ernst. 1997. *This Is Biology.* Cambridge, Mass: Belknap Press of Harvard University Press.

McGrath, Alister. 2005. "A Blast from the Past? The Boyle Lectures and Natural Theology," *Science and Christian Belief* 17:25-34.

McMullin, Ernan. 1981. "How Should Cosmology Relate to Theology?" In *The Sciences and Theology in the Twentieth Century,* edited by A. R. Peacocke, 17-57. Notre Dame: University of Notre Dame Press.

Merchant, Carolyn. 1980. *The Death of Nature: Women, Ecology, and the Scientific Revolution.* San Francisco: Harper & Row.

Miller, Kenneth R. 1999. *Finding Darwin's God: A Scientist's Search for Common Ground between God and Evolution.* New York: Cliff Street Books.

Moltmann, Jürgen. 1967. *Theology of Hope: On the Ground and the Implications of a Christian Eschatology.* Translated by James W. Leitch. New York: Harper & Row.

———. 1974. *The Crucified God: The Cross of Christ as the Foundation and Criticism of Christian Theology.* Translated by R. A. Wilson and John Bowden. New York: Harper & Row.

———. 1975. *The Experiment Hope.* Edited and translated by M. Douglas Meeks. Philadelphia: Fortress.

———. 1985. *God in Creation: A New Theology of Creation and the Spirit of God.* Translated by Margaret Kohl. San Francisco: Harper & Row.

———. 1992. *The Spirit of Life: A Universal Affirmation.* Translated by Margaret Kohl. Minneapolis: Fortress.

————. 1996. *The Coming of God: Christian Eschatology.* Translated by Margaret Kohl. Minneapolis: Fortress.

Monod, Jacques. 1971. *Chance and Necessity: An Essay on the Natural Philosophy of Modern Biology.* Translated by Austryn Wainhouse. New York: Knopf.

Mooney, Christopher. 1996. *Theology and Scientific Knowledge.* Notre Dame: University of Notre Dame Press.

Morowitz, Harold J. 1997. *The Kindly Dr. Guillotin and Other Essays on Science and Life.* Washington, D.C.: Counterpoint.

————. 2002. *The Emergence of Everything: How the World Became Complex.* New York: Oxford University Press.

Murchie, Guy. 1978. *The Seven Mysteries of Life: An Exploration in Science and Philosophy.* Boston: Houghton Mifflin.

Needleman, Jacob. 1976. *A Sense of the Cosmos.* New York: E. P. Dutton.

Newman, John Henry. 1959 [1854]. *The Idea of a University.* Garden City, N.Y.: Image Books.

Niebuhr, H. Richard. 1960. *The Meaning of Revelation.* New York: Macmillan.

Northrop, F. S. C. 1979. *Science and First Principles.* Woodbridge, Conn.: Ox Bow Press.

Ogden, Schubert. 1977. *The Reality of God and Other Essays.* New York: Harper & Row.

————. 1986. *On Theology.* San Francisco: Harper & Row.

Pagels, Heinz. 1985. *Perfect Symmetry.* New York: Bantam Books.

Paley, William. 2006 [1816]. *Natural Theology; or, Evidence of the Existence and Attributes of the Deity Collected from the Appearances of Nature.* Edited with an introduction by Matthew D. Eddy and David Knight. Oxford: Oxford University Press.

Pannenberg, Wolfhart. 1970. *What Is Man?* Translated by Duane A. Priebe. Philadelphia: Fortress.

————. 1977. *Faith and Reality.* Translated by John Maxwell. Philadelphia: Westminster.

————. 1993. *Toward a Theology of Nature: Essays on Science and Faith.* Edited by Ted Peters. Louisville: Westminster John Knox.

Papineau, David. 1993. *Philosophical Naturalism.* Cambridge, Mass.: Blackwell.

Peters, Ted. 2000. *God—The World's Future: Systematic Theology for a New Era.* 2nd ed. Minneapolis: Fortress.

Phipps, William E. 2002. *Darwin's Religious Odyssey.* Harrisburg, Penn.: Trinity Press International.

Polanyi, Michael. 1963. *The Study of Man.* Chicago: University of Chicago Press.

————. 1964. *Personal Knowledge: Towards a Post-Critical Philosophy.* New York: Harper Torchbooks.

————. 1967. *The Tacit Dimension.* Garden City, N.Y.: Doubleday Anchor Books.

————. 1969. *Knowing and Being.* Edited by Marjorie Grene. Chicago: University of Chicago Press.

Polanyi, Michael and Harry Prosch. 1975. *Meaning.* Chicago: University of Chicago Press.

Polkinghorne, John. 2000. *Quarks, Chaos and Christianity: Questions to Science and Religion.* New York: Crossroad.

Pollack, Robert E. 2000. *The Faith of Biology and the Biology of Faith.* New York: Columbia University Press.

Rahner, Karl. 1969. *Theological Investigations.* Vol. 6. Translated by Karl and Boniface Kruger. Baltimore: Helicon.

———. 1978. *Foundations of Christian Faith.* Translated by William V. Dych. New York: Crossroad.

Raymo, Chet. 1998. *Skeptics and True Believers: The Exhilarating Connection between Science and Religion.* New York: Walker.

Rees, Martin. 2000. *Just Six Numbers: The Deep Forces that Shape the Universe.* New York: Basic Books.

———. 2001. *Our Cosmic Habitat.* Princeton: Princeton University Press.

Richard, Lucien J., O.M.I. 1982. *A Kenotic Christology: In the Humanity of Jesus the Christ, the Compassion of Our God.* Lanham, Md.: University Press of America.

Ricoeur, Paul. 1965. *History and Truth.* Translated by Charles Kelbley. Evanston, Ill.: Northwestern University Press.

———. 1969. *The Symbolism of Evil.* Translated by Emerson Buchanan. Boston: Beacon.

———. 1974. *The Conflict of Interpretations: Essays in Hermeneutics.* Edited by Don Ihde. Evanston, Ill.: Northwestern University Press.

Robinson, John A. T. 1973. *The Human Face of God.* Philadelphia: Westminster.

Rolston, Holmes, III. 1987. *Science and Religion: A Critical Survey.* New York: Random House.

———. 1999. *Genes, Genesis and God: Values and Their Origins in Natural and Human History.* New York: Cambridge University Press.

Rose, Michael R. 1998. *Darwin's Spectre: Evolutionary Biology in the Modern World.* Princeton: Princeton University Press.

Rue, Loyal. 1994. *By the Grace of Guile: The Role of Deception in Natural History and Human Affairs.* New York: Oxford University Press.

Ruse, Michael. 2003. *Darwin and Design: Does Evolution Have a Purpose?* Cambridge, Mass.: Harvard University Press.

Russell, Bertrand. 1918. *Mysticism and Logic and Other Essays.* New York: Longmans, Green.

Russell, Robert John, Nancey Murphy, and C. J. Isham, eds. 1997. *Quantum Cosmology and the Laws of Nature.* Notre Dame: Vatican Observatory and University of Notre Dame Press.

Sagan, Carl. 1985. *Cosmos.* New York: Ballantine Books.

———. 1995. *The Demon-Haunted World: Science as a Candle in the Dark.* New York: Random House.

Schillebeeckx, Edward. 1965. *Christ, the Sacrament of Encounter with God.* Translated by Paul Barrett and N. D. Smith. New York: Sheed & Ward.

———. 1990. *Church: The Human Story of God.* Translated by John Bowden. New York: Crossroad.

Schleiermacher, Friedrich. 1958 [1799]. *On Religion: Speeches to Its Cultured Despisers.* New York: Harper & Row.

Schumacher, E. F. 1978. *A Guide for the Perplexed.* New York: Harper Colophon Books.

Shermer, Michael. 1997. *Why People Believe Weird Things: Pseudoscience, Superstition, and Other Confusions of Our Time.* New York: W. H. Freeman.

————. 2000. *How We Believe: The Search for God in an Age of Science.* New York: W. H. Freeman.

Simpson, George Gaylord. 1971. *The Meaning of Evolution.* Rev. ed. New York: Bantam Books.

Skinner, B. F. 1972. *Beyond Freedom and Dignity.* New York: Bantam Books.

Smith, Huston. 1976. *Forgotten Truth: The Primordial Tradition.* New York: Harper & Row.

————. 1982. *Beyond the Post-Modern Mind.* New York: Crossroad.

Smolin, Lee. 1997. *The Life of the Cosmos.* New York: Oxford University Press.

Stillman, Drake. 1957. *Discoveries and Opinions of Galileo.* New York: Anchor Books,.

Teilhard de Chardin, Pierre. 1960. *The Divine Milieu: An Essay on the Interior Life.* New York: Harper & Row.

————. 1962. *Human Energy.* Translated by J. M. Cohen. New York: Harvest Books/Harcourt Brace Jovanovich.

————. 1962. *Letters from a Traveler.* New York: Harper & Row.

————. 1964. *The Future of Man.* Translated by Norman Denny. New York: Harper & Row.

————. 1965. *Science and Christ.* Translated by René Hague. New York: Harper & Row.

————. 1966. *The Vision of the Past.* Translated by J. M. Cohen. New York: Harper & Row.

————. 1968. *Writings in Time of War.* Translated by René Hague. New York: Harper & Row.

————. 1969. *Hymn of the Universe.* Translated by Gerald Vann. New York: Harper Colophon.

————. 1969. *Christianity and Evolution.* Translated by René Hague. New York: Harcourt Brace Jovanovich.

————. 1969. *How I Believe.* Translated by René Hague. New York: Harper & Row.

————. 1970. *Activation of Energy.* Translated by René Hague. New York: Harcourt Brace Jovanovich.

————. 1999 [1959]. *The Human Phenomenon.* Translated by Sarah Appleton-Weber. Portland, Oregon: Sussex Academic Press.

Theissen, Gerd. 1991. *The Open Door: Variations on Biblical Themes.* Translated by John Bowden. Minneapolis: Fortress.

Thiel, John. 2002. *God, Evil, and Innocent Suffering: A Theological Reflection.* New York: Crossroad.

Thiemann, Ronald. 1985. *Revelation and Theology.* Notre Dame: University of Notre Dame Press.

Tillich, Paul. 1952. *The Courage to Be.* New Haven: Yale University Press.

———. 1958. *Dynamics of Faith.* New York: Harper Torchbooks.

———. 1963. *Systematic Theology.* 3 volumes. Chicago: University of Chicago Press.

———. 1996 [1948]. *The Shaking of the Foundations.* New York: Charles Scribner's Sons.

Toulmin, Stephen. 1970. *An Examination of the Place of Reason in Ethics.* Cambridge: Cambridge University Press.

Toulmin, Stephen, and June Goodfield. 1965. *The Discovery of Time.* London: Hutchinson.

Towers, Bernard. 1969. *Concerning Teilhard, and Other Writings on Science and Religion.* London: Collins.

Tracy, David. 1975. *Blessed Rage for Order: The New Pluralism in Theology.* New York: Seabury Press.

von Balthasar, Hans Urs. 1990. *Mysterium Paschale: The Mystery of Easter.* Translated by Aidan Nichols, O.P. Edinburgh: T & T Clark.

von Rad, Gerhard. *Old Testament Theology.* 1962-65. 2 vols. Translated by D. M. G. Stalker. New York: Harper & Row.

von Weizsäcker, Carl Friedrich. 1949. *The History of Nature.* Chicago: University of Chicago Press.

Wallace, B. Alan. 2000. *The Taboo of Subjectivity: Toward a New Science of Consciousness.* New York: Oxford University Press.

Watson, J. D. 1965. *The Molecular Biology of the Gene.* New York: W. A. Benjamin.

Weinberg, Steven. 1992. *Dreams of a Final Theory.* New York: Pantheon.

Wells, Jonathan. 2000. *Icons of Evolution: Science or Myth? Why Much of What We Teach about Evolution Is Wrong.* Washington, D.C.: Regnery.

White, Andrew Dickson. 1965. *A History of the Warfare of Science with Theology in Christendom.* New York: Free Press.

Whitehead, Alfred North. 1925. *Science and the Modern World.* New York: Free Press.

———. 1967. *Adventures of Ideas.* New York: Free Press.

———. 1968 [1938]. *Modes of Thought.* New York: Free Press.

———. 1968. *Process and Reality.* Corrected ed. Edited by David Ray Griffin and Donald W. Sherburne. New York: Free Press.

Wilson, A. N. 1999. *God's Funeral.* New York: W. W. Norton.

Wilson, E. O. 1998. *Consilience: The Unity of Knowledge.* New York: Knopf.

Index